QuickStudy®

for

Calculus

BarCharts inc
publishing

Boca Raton, Florida

©2007 BarCharts, Inc.

ISBN 13: 978-142320268-4

ISBN 10: 142320268-6

BarCharts® and QuickStudy® are registered trademarks of BarCharts, Inc.

Author: Dr. S. B. Kizlik

Publisher:

BarCharts, Inc.

6000 Park of Commerce Boulevard, Suite D

Boca Raton, FL 33487

www.quickstudy.com

Printed in Thailand

Contents

- Derivative Formulas
  - Constants
  - Reciprocal Function
  - Square Root
  - Powers
  - Exponentials
  - Logarithms
  - Hyperbolic Functions
    + Sine
    + Cosine
    + Arcsine
    + Arccosine
  - Trig Functions
    + Sine
    + Cosine
    + Tangent
    + Cotangent
    + Secant
    + Cosecant
    + Arcsine
    + Arccosine
    + Arctangent
    + Arccotangent

- *Notes*
- Local Features of Functions
  - Neighborhoods
  - Continuity
  - Critical Points
  - Local Extrema
    + Local Minimum Point
    + Local Maximum Point
    + Relative Extrema
  - First Derivative Test

Study Hints

NOTE TO STUDENT:

Use this QuickStudy® booklet to make the most of your studying time.

All equations are set in boldface type for easy reference.

EX: if P increases 4% each half year, then $a^{\frac{1}{2}} = 1.04$, and $P = P_0(1.04)^{2t} \approx P_0 e^{0.078t}$ (t in yrs).

Diagrams are presented in full color.

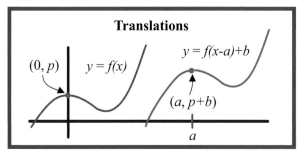

QuickStudy® notes provide chapter overviews; refer to them for easy reference.

NOTES
This chapter explains the basics of **functions** to **beginning calculus** students.

Take your learning to the next level with QuickStudy®!

1 Functions

NOTES
This chapter explains the basics of **functions** to **beginning calculus** students.

Definitions

■ **Function:** A correspondence that assigns one value (output) to each member of a given set.
  ◆ The given set of inputs is called the **domain**.
  ◆ The set of outputs is called the **range**.
  ◆ One-variable calculus deals with real-valued functions whose domain is a set of real numbers. If a domain is not specified, it is assumed to include all inputs for which there is a real number output.

■ **Notation:** If a function is named f, then $f(x)$ denotes its value at x, or "f evaluated at x."
  ◆ If a function gives a quantity y in terms of a variable quantity x, then x is called the **independent variable** and y the **dependent variable**.
  ◆ Given a function by an equation such as $y = x^2$, one may think of y as shorthand for the function's expression. The notation $x \mapsto x^2$ ("x maps to x^2") is another way to refer to the function.

◆ The *expression f(x)* for a function at an arbitrary input x often stands in for the function itself.

■ **Graph:** The graph of a function f is the set of ordered pairs $(x, f(x))$, presented visually on a Cartesian coordinate system.

◆ The **vertical line test** states that a curve is the graph of a function if every vertical line intersects the curve at most once.

◆ An equation $y = f(x)$ often refers to the set of points (x, y) satisfying the equation, in this case the graph of the function f.

◆ The **zeros** of a function are the inputs x for which $f(x) = 0$, and they are the **x-intercepts** of the graph.

■ **Even & Odd:** A function f is **even** if $f(-x) = f(x)$ and **odd** if $f(-x) = -f(x)$. Most are neither.

Numbers

■ **Rational Numbers:** A rational number is a ratio p/q of integers p and q, with $q \neq 0$.

◆ There are infinite ways to represent a given rational number, but there is a unique lowest-terms representative.

◆ The set of all rational numbers forms a closed system under the usual arithmetic operations; except division by zero.

■ **Real Numbers:** In this guide, **R** denotes the set of real numbers. Real numbers may be thought of as the numbers represented by infinite decimal expansions.

◆ **Rational numbers** terminate in all zeros or have a repeating segment of digits.

◆ **Irrational numbers** are real numbers that are not rational.

EX: π is irrational; it may be *approximated* by rational numbers such as $^{22}/_7$ or **3.1416**.

■ **Machine Numbers:** A calculator or computer approximately represents real numbers using a fixed number of digits, usually between 8 and 16. Machine calculations are therefore usually not exact. This can cause anomalies in plots.

◆ The **precision** of a numerical result is the number of correct digits. (Count digits after appropriate rounding: 2.512 for 2.4833 has two correct digits.)

◆ The **accuracy** refers to the number of correct digits after the decimal point.

■ **Intervals:** If $a < b$, the **open interval (a,b)** is the set of real numbers x such that $a < x < b$.

◆ The **closed interval $[a,b]$** is the set of x such that $a \leq x \leq b$.

◆ The notation $(-\infty, a)$ denotes the half-line consisting of all real numbers x such that $x < a$ (or $-\infty < x < a$). Likewise, there are intervals of the form $(-\infty, a)$ and (a, ∞).

◆ The symbol ∞ is not to be thought of as a number. The entire real number line is an interval, $\mathbf{R} = (-\infty, \infty)$.

New Functions

■ **Arithmetic:** The **scalar multiple** of a function f by a constant c is given by $(cf)(x) = c \cdot f(x)$. The sum $f + g$, product fg, and quotient f/g of functions f and g are defined by:

$$(f+g)(x) = f(x) + g(x),$$
$$(fg)(x) = f(x)g(x),$$
$$(f/g)(x) = f(x)/g(x)$$

♦ In each case, the domain of the new function is the intersection of the domains of f and g, with the zeros of g excluded for the quotient.

■ **Composition:** If f and g are functions, "f composed with g" is the function $f \circ g$ given by $(f \circ g)(x) = f(g(x))$ with domain (strictly speaking) the set of x in the domain of g for which $g(x)$ is in the domain of f.

■ **Translations:** The graph of $x \mapsto (x-a)$ is the graph of $f(x)$ translated a units to the right; the graph of $x \mapsto f(x)+b$ is the graph of f translated b units upward.

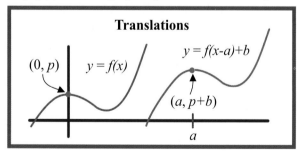

Translations

$(0, p)$ $y = f(x)$

$y = f(x-a)+b$

$(a, p+b)$

a

■ **Inverses:** An **inverse** of a function f is a function g or f^{-1} such that $g(f(x)) = f^{-1}(f(x)) = x$ for all x in the domain of f.

◆ A function f has an inverse if and only if it is **one-to-one**: for each of its values y there is only one corresponding input; or, $f(x) = y$ has only one solution; or, any horizontal line intersects the graph of f at most once.

EX: x^3 is one-to-one, x^2 is not. Strictly increasing or decreasing functions are one-to-one.

◆ There can be only one inverse defined on the range of f, denoted f^{-1}. For any y in the range of f, $f^{-1}(y)$ is the x that solves $f(x) = y$.

◆ If the axes have the same scale, the graph of f^{-1} is the reflection of the graph of f across the line $y = x$.

■ **Implicit Functions:** A relation $F(x,y) = c$ often admits y as a function of x, in one or more ways.

EX: $x^2 + y^2 = 4$ admits $y = \sqrt{4-x^2}$. Such functions are said to be implicitly defined by the relation. Graphically, the relation gives a curve, and a piece of the curve satisfying the vertical line test is the graph of an implicit function. Often, there is no expression for an implicit function in terms of elementary functions.

EX: $x^2\, 2y + y^2\, 2x = 4$ admits $y = f(x)$, but there is no formula for $f(x)$.

Elementary Algebraic Functions
■ Constant & Identity:
A constant function has only one output: $f(x) = c$.
The identity function is: $x| \rightarrow x$, or $f(x) = x$.

■ **Absolute Value:** $|x| = \begin{cases} -x \text{ if } x < 0 \\ x \text{ if } x \geq 0 \end{cases}$

The above is an example of a **piecewise definition.**

For any x, $\sqrt{x^2} = |x|$.

■ **Linear Functions:** For a linear function, the difference of two outputs is proportional to the difference of inputs.

♦ The proportionality constant, i.e., the ratio of output difference to input difference $m = \frac{y_2 - y_1}{x_2 - x_1}$ is called the **slope.**

♦ The slope is also the change in the function per unit increase in the independent variable.

♦ The linear function $f(x) = mx + b$ has slope m and **y-intercept** $f(0) = b$, the graph is a straight line.

♦ The linear function with value y_0 at x_0 and slope m is $f(x) = y_0 + m(x - x_0)$.

■ **Quadratics:** These have the form $f(x) = ax^2 + bx + c$ where $(a \neq 0)$.

♦ The **normal form** is $f(x) = a(x-h)^2 + k$.

♦ Also, $h = -b/(2a)$ and $k = f(h)$.

♦ The graph is a parabola with **vertex (h,k),** opening up or down accordingly as $a > 0$ or $a < 0$, and symmetric about the vertical line through vertex.

♦ A quadratic has two, one, or no zeros accordingly as the **discriminant $b^2 - 4ac$** is positive, zero, or negative.

♦ Zeros are given by the **quadratic formula**

$$x = \frac{-b \pm \sqrt{b^2 - 4ac}}{2a}$$

and are graphically located symmetrically on either side of the vertex.

■ **Polynomials:** These have the form

$$p(x) = ax^n + bx^{n-1} + ... + dx + e.$$

◆ Assuming $a \neq 0$, this has **degree** n, **leading coefficient** a, and **constant term** $e = p(0)$.

◆ A polynomial of degree n has at most n zeros.

◆ If x_0 is a zero of $p(x)$, then $x - x_0$ is a factor of $p(x)$: $p(x) = (x - x_0) \, q(x)$

for some degree n-1 polynomial $q(x)$.

◆ A polynomial graph is smooth and goes to $\pm \infty$ when $|x|$ is large.

■ **Rational Functions:** These have the form $f(x) = \dfrac{p(x)}{q(x)}$ where $p(x)$ and $q(x)$ are polynomials.

◆ The domain excludes the zeros of q.

◆ The zeros of f are the zeros of p that are not zeros of q.

◆ The graph of a rational function may have vertical asymptotes and removable discontinuities, and is similar to some polynomial (perhaps constant) when $|x|$ is large.

■ *n*th **Roots:** These have the form $y = x^{\frac{1}{n}} \equiv \sqrt[n]{x}$ for some integer $n > 1$.

◆ If n is even, the domain is $(0, \infty)$ and y is the unique nonnegative number such that $y^n = x$.

◆ If n is odd, the domain is $\mathbf{R}$ and y is the unique real number such that $y^n = x$.

■ **Algebraic vs. Transcendental:** An algebraic function $y = f(x)$ satisfies a two-variable polynomial equation $P(x,y) = 0$. The functions above are algebraic.

EX: $y = |x|$ satisfies $x^2 - y^2 = 0$. Sums, products, quotients, powers, and roots of algebraic functions are **algebraic**. Functions that are not algebraic (e.g., exponentials, logarithms, and trig functions) are called **transcendental**.

■ **Rational Powers:** These have the form $f(x) = x^{\frac{m}{n}} \equiv (x^m)^{\frac{1}{n}} \equiv (x^{\frac{1}{n}})^m$ where it is assumed m and n are integers, $n > 0$, and $|m|/n$ is in lowest terms.

◆ If $m < 0$ then $x^m = 1/x^{|m|}$. The domain of $x^{m/n}$ is the same as that of the nth root function, excluding 0 if $m < 0$.

◆ For $x > 0$, as p decreases in absolute value, graphs of $y = x^p$ move toward the line $y = x^0 = 1$; as p increases in absolute value, graphs of $y = x^p$ move away from the line $y = 1$ and toward the line $x = 1$.

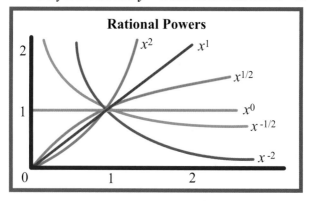

Rational Powers

Exponentials & Logarithms

■ **Pure Exponentials:** The pure exponential function with **base** a $(a > 0,\ a \neq 1)$ is $f(x) = a^x$.

◆ The domain is **R** and the range is $(0, \infty)$. The y-intercept is $a^0 = 1$. If $a < 1$, the function is decreasing; if $a > 1$, it is increasing. It changes by the factor $a^{\Delta x}$ over any interval of length Δx.

◆ Exponentials turn *addition into multiplication*:
$$a^0 = 1$$
$$a^{x+y} = a^x a^y$$
$$a^{mx} = (a^x)^m$$
$$a^{x-y} = a^x/a^y$$
$$a^{-x} = 1/a^x$$

■ **Logarithms:** The logarithm with base a is the inverse of the base a exponential:
$\log_a x = $ "the power of a that yields x"; thus, $\log_a x = y$ if $a^y = x$

◆ Equivalently, $x = a^{\log_a x}$ or $\log_a a^y = y$.

◆ The domain of $\log_a$ is $(0, \infty)$ and the range is **R**. If $a > 1$, then $\log_a x$ is negative for $0 < x < 1$, positive for $x > 1$, and always increasing. The **common logarithm** is $\log_{10}$.

   EX:
$$\log_a a = 1$$
$$\log_2 32 = 5,$$
$$\log_{10} (1/10) = -1$$

◆ Logarithms turn *multiplication into addition*:
$$\log_a 1 = 0$$
$$\log_a xy = \log_a x + \log_a y$$
$$\log_a x^m = m\log_a x$$
$$\log_a (x/y) = \log_a x - \log_a y$$
$$\log_a (1/x) = -\log_a x$$

◆ The third identity holds for any real number m.

   For a **change of base**, one has $\log_b x = \dfrac{\log_a x}{\log_a b}$

■ **Natural Exponential & Logarithm:** The natural exponential function is the pure exponential whose tangent line at the point **(0, 1)** on its graph has slope of **1**. Its base is an irrational number:

$$e = \lim_{n \to \infty}\left(1+\frac{1}{n}\right)^n \approx 2.718.$$

◆ The **natural logarithm** is $\ln = \log_e$, the inverse to $x|\!\to\! e^x$; $\ln x = y$ means $x = e^y$.

◆ There are identities

$$\ln e^x = x,$$
$$e^{\ln x} = x,$$
$$\ln e = 1,$$

and **ln** has the properties of a logarithm.

EX: $\ln(1/x) = -\ln x$. Special values are: $\ln 1 = 0$, $\ln 2 \approx 0.6931$, $\ln 10 \approx 2.303$.

◆ Any exponential can be written $a^x = e^{(\ln a)x}$.

◆ Any logarithm can be written $\log_a x = \dfrac{\ln x}{\ln a}$.

■ **General Exponential Functions:** These have the form $f(x) = P_0 a^x$ and have the property that the ratio of two outputs depends only on the difference of inputs.

◆ The ratio of outputs for a unit change in input is the **base** a.

◆ The y-intercept is $f(0) = P_0$.

■ **Exponential Growth:** A quantity **P** (e.g., invested money) that increases by a factor $a = e^r > 1$ over each unit of time is described by $P = P_0 a^t = P_0 e^{rt}$.

◆ Over an interval Δt the factor is $a^{\Delta t}$.

EX: if P increases 4% each half year, then $a^{\frac{1}{2}} = 1.04$, and $P = P_0(1.04)^{2t} \approx P_0 e^{0.078t}$ (t in yrs).

◆ The **doubling time** D is the time interval over which the quantity doubles: $a^D = e^{rD} = 2$,
$$D = \frac{\ln 2}{\ln a} = \frac{\ln 2}{r}$$

◆ If the doubling time is D, then $P = P_0 2^{t/D}$.

■ **Continuous Compounding:** At the annual percentage rate $r \times 100\%$ yields the annual growth factor $a = \lim_{n \to \infty}\left(1 + \frac{r}{n}\right)^n e^r$; also, $a = Pe^{rt}$

■ **Exponential Decay:** A quantity Q (e.g., of radioactive material) that decreases to a proportion $b = e^{-k} < 1$ over each unit of time is described by $Q = Q_0 b^t = Q_0 e^{-kt}$.

◆ Over an interval Δt the proportion is $b^{\Delta t}$.

EX: if Q decreases 10% every 12 hours, then $b^{12} = 0.90$, and $Q = Q_0(0.90)^{t/12} \approx Q_0 e^{-.0088t}$ (t in hrs).

◆ The **half-life** H is the time interval over which the quantity decreases by the factor one-half: $b^H = e^{-kH} = 1/2$, $H = \frac{\ln 2}{\ln b} = \frac{\ln 2}{k}$.

◆ If the half-life is H, then $Q = Q_0(1/2)^{t/H}$.

■ **Irrational Powers:** These may be defined by $f(x) = x^p = e^{p \ln x}$ where $(x > 0)$.

■ **Hyperbolic Functions:** The **hyperbolic cosine** is $\cosh x = \frac{e^x + e^{-x}}{2}$.

◆ It has domain **R**, range $[1,\infty)$, and is even.

◆ On the restricted domain $[0,\infty)$, it has inverse **arccosh** $x \equiv \cosh^{-1}x = \ln\left(x + \sqrt{x^2 - 1}\right)$.

◆ The **hyperbolic sine** is $\sinh x = \frac{e^x - e^{-x}}{2}$.

◆ It has domain **R**, range **R**, and is odd. Always strictly increasing, it has inverse **arcsinh** $x \equiv$ **sinh**$^{-1}x = $ **ln** $\left(x+\sqrt{x^2+1}\right)$.

◆ The basic identity is **cosh**^2x **- sinh**^2x **= 1**.

Trigonometric Functions

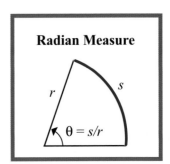

Radian Measure

■ **Radians:** The radian measure of an angle θ is the ratio of length s to radius r of a corresponding circular arc: $\theta = \frac{s}{r}$.

$$\theta = s/r$$

$$2\pi \text{ radians} = 360°$$

$$1° = \frac{\pi}{180}$$

◆ In calculus, it is normally assumed (and necessary for standard derivative formulas) that arguments to trigonometric (trig) functions are in radians.

■ **Cosine, Sine & Tangent:** Consider a real number t as the radian measure of an angle: the distance measured counter-clockwise along the circumference of the *unit* circle from the point **(1, 0)** to a terminal point **(x, y)**.

◆ Then **cos** $t = x$; **sin** $t = y$; **tan** $t = \dfrac{\sin t}{\cos t} = \dfrac{y}{x}$

◆ Cosine and sine have domain **R** and range **[-1,1]**.

◆ The domain of the tangent excludes $\pm\frac{1}{2}\pi$, $\pm\frac{3}{2}\pi$ ... , and its range is **R**.

◆ The cosine is even, the sine and tangent are odd.

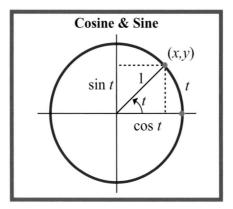

Cosine & Sine

■ **Secant, Cosecant & Cotangent:**

$$\sec t = \frac{1}{\cos t}; \ \csc t = \frac{1}{\sin t}; \ \cot t = \frac{1}{\tan t} = \frac{\cos t}{\sin t}$$

■ **Special Values:**

t	0	$\frac{\pi}{6}$	$\frac{\pi}{4}$	$\frac{\pi}{3}$	$\frac{\pi}{2}$	π
cos t	1	$\frac{\sqrt{3}}{2}$	$\frac{\sqrt{2}}{2}$	$\frac{1}{2}$	0	-1
sin t	0	$\frac{1}{2}$	$\frac{\sqrt{2}}{2}$	$\frac{\sqrt{3}}{2}$	1	0
tan t	0	$\frac{\sqrt{3}}{3}$	1	$\sqrt{3}$	Ø	0

■ **Identities:**

$$\sin^2 t + \cos^2 t = 1$$
$$\tan^2 t + 1 = \sec^2 t$$
$$\sin(a + b) = \sin a \cos b + \cos a \sin b$$
$$\cos(a + b) = \cos a \cos b - \sin a \sin b$$

◆ Other identities are obtained from the above.

EX: $\sin(t - \pi/2) = -\cos t$

$$\cos(2t) = \cos^2 t - \sin^2 t = 1 - 2\sin^2 t$$

$$\tan(a + b) \ \frac{\tan a + \tan b}{1 - \tan a \tan b}$$

■ **Amplitudes, Periods & Phases:** If $f(t) = A\sin(\omega t + \phi) + k$ with $A > 0$, $\omega > 0$, and $-\pi < \phi \le \pi$, then the **amplitude** is A, the **average value** is k, the **period** is $\frac{2\pi}{\omega}$, the **frequency** $\left(\frac{1}{\text{period}}\right)$ is $\frac{\omega}{2\pi}$, the **angular frequency** is ω, and the **phase shift** (relative to $A\sin \omega t$) is ϕ.

■ **Inverse Trig Functions:**

◆ The **arccosine** is inverse to cosine on $[0,\pi]$: **arccos** x = "angle in $[0,\pi]$ whose cosine is x." It has domain $[-1,1]$ and range $[0,\pi]$.

$$\arccos\left(\frac{\sqrt{3}}{2}\right) = \frac{\pi}{6}, \ \arccos\left(\frac{-\sqrt{2}}{2}\right) = \frac{3\pi}{4}$$

◆ The **arcsine** is inverse to sine on: $\left[\frac{-\pi}{2}, \frac{\pi}{2}\right]$ **arcsin** x = "angle in $\left[\frac{-\pi}{2}, \frac{\pi}{2}\right]$ whose sine is x." It has domain $[-1,1]$, range $\left[\frac{-\pi}{2}, \frac{\pi}{2}\right]$, and is odd.

◆ The **arctangent** is inverse to tangent on: $\left(\frac{-\pi}{2}, \frac{\pi}{2}\right)$ **arctan** x = "angle in $\left(\frac{-\pi}{2}, \frac{\pi}{2}\right)$ with tangent x." It has domain **R**, range $\left(\frac{-\pi}{2}, \frac{\pi}{2}\right)$, and is odd.

$$\arctan \sqrt{3} = \frac{\pi}{3}, \ \arctan(-1) = \frac{-\pi}{4}$$

◆ The notation $\cos^{-1} x$ for **arccos** x is not to be confused with $1/\cos x$; likewise $\sin^{-1} x$ and $\tan^{-1} x$.

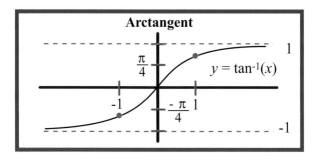

2 Limits

> **NOTES**
> This chapter explains the basics of **limits** to **beginning calculus** students.

Definitions

■ **Limit:** Intuitively, the limit of $f(x)$ as x approaches a is the number that $f(x)$ gets close to when x gets close to a.

◆ Precisely, a number L is the limit, written $\lim\limits_{x \to a} f(x) = L$ or $f(x) \mid\to L$ as $x \to a$ if every $\varepsilon > 0$ admits a $\delta > 0$ such that $|f(x)-L| < \varepsilon$ when $0 < |x-a| < \delta$

◆ It is assumed that $f(x)$ is defined for all x in some open interval containing a, except perhaps $x = a$.

◆ If a limit exists, there is only one. The limit statement says nothing whatever about the value of f at $x = a$.

■ **Zooming Formulation:** If the plot range for f is held fixed with L in the middle, and the plot domain is narrowed through intervals centered at $x = a$, the graph of f eventually lies completely within the fixed plot range, except perhaps at $x = a$.

■ **One-Sided Limits:**

◆ The left-hand limit is equal to **L**, written $\lim_{x \to a^-} f(x) = L$ or $f(a^-) = L$, if every $\varepsilon > 0$ admits a $\delta > 0$ such that $|f(x) - L| < \varepsilon$ when $a - \delta < x < a$.

◆ The right-hand limit is defined similarly, the last condition being $a < x < a + \delta$.

EX: $\lim_{x \to 0^+} \arctan\left(\frac{1}{x}\right) = \frac{\pi}{2}$.

◆ A limit exists if and only if the left and right-hand limits exist and are equal.

■ **Infinite Limits:** One writes $\lim_{x \to a} f(x) = \infty$ if every $Y > 0$ admits a $\delta > 0$ such that $f(x) > Y$ when $0 < |x - a| < \delta$.

◆ Likewise, there are one-sided limits to ∞, and limits to $-\infty$.

EX: $\lim_{x \to 0^-} \left(\frac{1}{x}\right) = -\infty$.

■ **Limits at Infinity:** One writes $\lim_{x \to \infty} f(x) = L$ if every $\varepsilon > 0$ admits an $X > 0$ such that $|f(x) - L| < \varepsilon$ when $x > X$.

Limit Theorems

(**Note:** The following theorems have counterparts involving limits to infinity. Also, "for x near a" will mean "for all x in some open interval containing a, except perhaps $x = a$.")

■ **Arithmetic:**

◆ A limit of a sum is the sum of the individual limits, provided each individual limit exists.

◆ Likewise for a limit of a difference or a product.

◆ The limit of a quotient is the quotient of the

individual limits, provided each individual limit exists and the limit of the denominator is nonzero.

◆ If c is a scalar, then $\lim\limits_{x \to a} cf(x) = c \lim\limits_{x \to a} f(x)$.

■ **Compositions:** If $\lim\limits_{x \to a} g(x) = l$ and $g(x) \neq l$ for x near a (or if F is continuous at l), then $\lim\limits_{x \to a} F(g(x)) = \lim\limits_{y \to l} F(y)$ provided the limit on the right exists.

EX: $\lim\limits_{x \to 1} (x^2+1)^n = \lim\limits_{y \to 2} y^n = 2^n$

$$\lim\limits_{x \to 0} \frac{\sin 5x}{5x} = \lim\limits_{y \to 0} \frac{\sin y}{y} = 1$$

■ **Inequalities:** If $f(x) \leq M$ for x near a, then $\lim\limits_{x \to a} f(x) \leq M$ if the limit exists. Likewise if $f(x) \geq m$.

■ **Sandwich Theorem:** If $g(x) \leq f(x) \leq h(x)$ for x near a, and $\lim\limits_{x \to a} g(x) = \lim\limits_{x \to a} h(x) = L$, then $\lim\limits_{x \to a} f(x) = L$.

Special case: if $|f(x)| \leq h(x)$ for x near a, then $\lim\limits_{x \to a} h(x) = 0$ implies $\lim\limits_{x \to a} f(x) = 0$

EX: $\lim\limits_{x \to 0} x \sin\left(\dfrac{1}{x}\right) = 0$ (using $h(x) = |x|$).

■ **L'Hôpital's Rule (requires derivatives):**

If $\lim\limits_{x \to a} f(x) = 0 = \lim\limits_{x \to a} g(x)$, and if $f'(x)$ and $g'(x)$ are defined and $g'(x) \neq 0$ for x near a, then $\lim\limits_{x \to a} \dfrac{f(x)}{g(x)} = \lim\limits_{x \to a} \dfrac{f'(x)}{g'(x)}$ provided the latter limit exists (or is infinite). The rule also holds when the limits of f and g are $\pm \infty$.

Limit Formulas

■ Polynomials & Rational Functions:

◆ If c is a constant, $\lim\limits_{x \to a} c = c$.

◆ If $p(x)$ is a polynomial, $\lim\limits_{x \to a} p(x) = p(a)$.

◆ Let $p(x)$ and $q(x)$ be polynomials.

• If $q(a) \neq 0$, then $\lim\limits_{x \to a} \dfrac{p(x)}{q(x)} = \dfrac{p(a)}{q(a)}$.

• If $q(a) = 0$ and $p(a) \neq 0$, one-sided limits are $\pm \infty$.
EX: For integer $n > 0$,

$$\lim\limits_{x \to 0^+} \frac{1}{x^n} = \infty;$$

$$\lim\limits_{x \to 0^-} \frac{1}{x^n} = -\infty \ (n \text{ odd});$$

$$\lim\limits_{x \to 0^-} \frac{1}{x^n} = \infty \ (n \text{ even}).$$

If $q(a) = 0$ and $p(a) = 0$, first cancel all common factors of $x - a$ from $p(x)$ and $q(x)$.

EX: $\lim\limits_{x \to 1^-} \dfrac{x^2 + 2x - 3}{x^3 - 3x + 2} = \lim\limits_{x \to 1^-} \dfrac{x+3}{(x-1)(x+2)}$

■ Rational Functions at Infinity:

For integer $n > 0$,

$$\lim\limits_{x \to \infty} x^n = \infty;$$

$$\lim\limits_{x \to -\infty} x^n = -\infty \ (n \text{ odd});$$

$$\lim\limits_{x \to -\infty} x^n = \infty \ (n \text{ even});$$

$$\lim\limits_{x \to \pm\infty} \frac{1}{x^n} = 0.$$

$$\lim\limits_{x \to \pm\infty} \frac{ax^n + bx^{n-1} + \dots}{cx^m + dx^{m-1} + \dots} = \frac{a}{c} \lim\limits_{x \to \pm\infty} x^{n-m} \ (a, c \text{ non zero})$$

■ **Arbitrary Powers:**

$\lim_{x \to a} x^p = a^p$ (when a^p is defined)

For $p > 0$, $\lim_{x \to \infty} x^p = \infty$ and $\lim_{x \to \infty} x^{-p} = 0$.

■ **Limits for Basic Derivatives:**

$\lim_{x \to a} \dfrac{x^m - a^m}{x - a} = ma^{m-1}$ (when a^{m-1} is defined)

$\lim_{h \to 0} \dfrac{e^h - 1}{h} = 1$ (a definition of e)

$\lim_{h \to 0} \dfrac{a^h - 1}{h} = \ln a$

$\lim_{x \to 0} \dfrac{\sin x}{x} = 1$

$\lim_{x \to 0} \dfrac{\cos x - 1}{x} = 0$

Continuity

Definitions

■ **Continuity at a Point:** A function f is continuous at a if a is in the domain of f and $\lim_{x \to a} f(x) = f(a)$.

 ◆ Explicitly, f is defined on some open interval containing a, and every $\varepsilon > 0$ admits a $\delta > 0$ such that $|f(x) - f(a)| < \varepsilon$ when $|x - a| < \delta$

■ **Zooming Formulation:** If the plot range for f is held fixed with $f(a)$ in the middle, and the plot domain is narrowed through intervals centered at $x = a$, the graph of f eventually lies completely within the fixed plot range. This must hold for any such plot range.

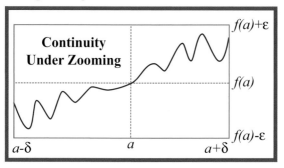

■ **One-Sided Continuity:** A function f is continuous from the left at a if a is in the domain of f and $\lim_{x \to a^-} f(x) = f(a)$.

  ◆ A function f is continuous from the right at a if a is in the domain of f and $\lim_{x \to a^+} f(x) = f(a)$.

■ **Global Continuity:** We say a function is continuous if it is continuous on its domain, meaning continuous at every point in its domain, using one-sided continuity at endpoints of intervals. (**Caution:** Textbooks sometimes refer to some points not in the domain as points of discontinuity. Intuitively, a function is continuous on an interval if there are no breaks in its graph.)

■ **Uniform Continuity:** A function f is uniformly continuous on its domain D if for every $\varepsilon > 0$ there is a $\delta > 0$ such that x, y in D and $|x - y| < \delta$ imply $|f(x) - f(y)| < \varepsilon$.

  ◆ Uniform continuity implies continuity. A continuous function on a closed interval $[a,b]$ is uniformly continuous.

Theory

■ **Arithmetic:** Scalar multiples of a continuous function are continuous. Sums, differences, products, and quotients of continuous functions are continuous (*on their domains*).

■ **Compositions:** A composition of continuous functions is continuous.

■ **Elementary Functions:** Polynomials, rational functions, root functions, exponentials and logarithms, and trigonometric and inverse trigonometric functions are continuous.

■ **Intermediate Value Theorem:** If f is continuous on the closed interval $[a,b]$, then f achieves every value between $f(a)$ and $f(b)$: for every y between $f(a)$ and $f(b)$ there is at least one x in $[a,b]$ such that $f(x) = y$.

◆ The **zero theorem** states that if f is continuous on $[a,b]$ and $f(a)$ and $f(b)$ have opposite signs, then there is an x in (a,b) such that $f(x) = 0$.

■ **Bisection Method:** This a method of finding zeros based on the zero theorem.

◆ With f, a, b as in the zero theorem, the midpoint $x_1 = \frac{1}{2}(a+b)$ is an initial estimate of a zero.

◆ Assuming $f(x_1)$ is nonzero, there is a new interval $[a, x_1]$ or $[x_1, b]$ on which opposite signs are taken at the endpoints. It contains a zero, and its midpoint x_2 is a new estimate of a zero.

◆ Repeat step (2) with the new interval and x_2.

◆ The nth estimate x_n differs from a zero by no more than $(b-a)/2^n$.

■ **Extreme Value Theorem:** If f is continuous on the closed interval $[a,b]$, then f achieves a minimum and a maximum on $[a,b]$: there are c and d in $[a,b]$ such that $f(c) \leq f(x) \leq f(d)$ for all x in $[a,b]$. (The proofs of this and the intermediate value theorem use properties of the set of real numbers not covered in introductory calculus.)

4

Derivatives

NOTES

This chapter explains the basics of **derivatives** to **beginning calculus** students.

Definitions

■ **Derivative:** The derivative of f at a is the number
$f'(a) = \lim_{h \to 0} \dfrac{f(a+h) - f(a)}{h}$ provided the limit
exists, in which case f is said to be **differentiable** at a.

◆ The derivative of f is the function f'.

◆ The derivative is also $f'(a) = \lim_{x \to a} \dfrac{f(x) - f(a)}{x - a}$
by the limit theorem for compositions applied to
$x | \to F(x - a)$, with $F(h) = \dfrac{f(a+h) - f(a)}{h}$

■ **Zooming Formulation:** If the plot domain for f is narrowed through intervals centered at $x = a$, while the ratio of the plot range to the plot domain is held fixed, the graph of f eventually appears linear (identical to the **tangent line** at $x = a$).

◆ If $f'(a) \neq 0$, the zoomed graph appears linear with no constraint on the plot ranges (auto-scaling).

■ **Notation:** The derivative function itself is denoted f' or $D(f)$.

◆ If $y = f(x)$, the following usually represent *expressions* for the derivative function:

$$y',\ \frac{dy}{dx},\ D_x y,\ f'(x),\ \frac{d}{dx}f(x)$$

◆ The second is the **Liebniz notation**. Notations for the derivative evaluated at $x = a$ are

$$f'(a),\ D(f)(a),\ \frac{dy}{dx}\bigg|_{x=a},\ \frac{dy}{dx}\bigg|_{x=a}f(x).$$

■ **Linearization:** The linearization, or **linear approximation**, of f at a is the linear function
$$x| \rightarrow f(a) + f'(a)(x-a)$$

◆ Its graph is the **tangent line** to the graph of f at the point $(a, f(a))$. The derivative thus provides a 'linear model' of the function near $x = a$.

■ **Differentials:** The differential of f at a is the expression $df(a) = f'(a)dx$.

◆ Applied to an increment Δx, it becomes $f'(a)\Delta x$.

◆ If $y = f(x)$, one writes $dy = f'(x)dx$.

■ **Difference Quotients:** The difference quotient $\dfrac{f(a+h)-f(a)}{h}$ approximates $f'(a)$ if h is small.

◆ It is the slope of the **secant line** through the points $(a, f(a))$ and $(a+h, f(a+h))$.

◆ The average of it and the "backward quotient,"
$\dfrac{f(a)-f(a-h)}{h}$ is the **symmetric quotient**

$\dfrac{f(a+h)-f(a-h)}{2h}$ usually a better approximation of $f'(a)$.

Interpretations

■ **Rate of Change:** The derivative $f'(a)$ is the **instantaneous rate of change** of f with respect to x at $x=a$. It tells how fast f is increasing or decreasing as x increases through values near $x = a$.

◆ The **average rate of change** of f over an interval $[a,x]$ is $\dfrac{f(x)-f(a)}{x-a}$. As x nears a, these average rates approach $f'(a)$.

◆ The **units of the derivative** are the units of $f(x)$ divided by the units of x.

■ **Tangent Line:** The derivative $f'(a)$ is the slope of the tangent line to the graph of f at the point $(a, f(a))$. It is a limit of slopes of secant lines passing through that point.

■ **Linear Approximation:** One can approximate values of f near a according to $f(x) \approx f(a) + f'(a)(x-a)$

EX: since, $\dfrac{d}{dx}\sqrt{x} = \dfrac{1}{2\sqrt{x}}$,

$$\sqrt{62} \approx \sqrt{64} + \frac{1}{2\sqrt{64}}(-2) = 7.875$$

◆ The approximation is better the closer x is to a and the flatter the graph is near a.

■ **Differential Changes:** At a given input, the derivative is the factor by which small input changes are scaled to become approximate output changes.

◆ The differential change at a over an input increment Δx approximates the output change:
$$f'(a)\Delta x \approx f(a + \Delta x) - f(a)$$

◆ The differential change is the exact change in the linear approximation.

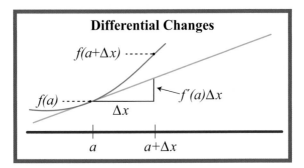

Differential Changes

- ■ **Velocity:** Suppose $s(t)$ is the position at time t of an object moving along a straight line.
  - ◆ Its **average velocity** over a time interval t_0 to t_1 is $\dfrac{s(t_1) - s(t_0)}{t_1 - t_0}$
  - ◆ Its **instantaneous velocity** at time t is $v(t) = s'(t)$.
  - ◆ Its **speed** is $|v(t)|$. Its **acceleration** is $v'(t)$.
- ■ **Interpreting a Derivative Value:** Suppose T is temperature (in °C) as a function of location x (in cm) along a line. The meaning of, for example, $T'(8) = 0.31$ (°C/cm) is, at the location $x = 8$, small shifts in the positive x direction yield small increases in temperature in a ratio of about 0.31 °C per cm shift. Small shifts in the negative direction yield like decreases in T.

Applications

- ■ **Linear Approximations at 0:**
  - ◆ The following are commonly used linear approximations valid near $x = 0$.
  $$\sin x \approx x$$
  $$\tan x \approx x$$
  $$e^x \approx 1 + x$$

$$\ln (1 + x) \approx x$$
$$(1 + x)^{1/2} \approx 1 + x/2$$
$$1/(1 + x) \approx 1 - x$$

◆ The error in each approximation is no more than $M|x|^2/2$, where M is any bound on $|f''(y)|$ for $|y| \le |x|$, f being the relevant function.

   EX: $|\sin x - x| \le .005$ for $|x| \le 0.1$.

■ **Newton's Method:** To find an approximate root of $f(x) = 0$, select an appropriate starting point x_0, and evaluate $x_{n+1} = x_n - f(x_n) / f'(x_n)$ successively for $n = 0, 1, ...,$ until the values do not change at the desired precision.

◆ The value on the right hand side in the above is where the tangent line at $(x_n, f(x_n))$ meets the **x-axis**.

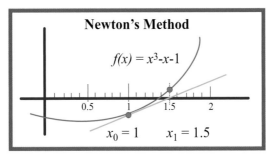

Newton's Method

$f(x) = x^3 - x - 1$

$x_0 = 1 \qquad x_1 = 1.5$

■ **Related Rates:** Suppose two variables, each a function of 'time,' are related by an equation.

◆ Differentiate both sides of the equation with respect to time to get a relation involving the time derivatives—the rates—and the original variables.

◆ With sufficient data for the variables and one of the rates, the derivative relation can be solved for the other rate.

Differentiation Rules

■ In the following, assume *f* and *g* are differentiable. Each rule should be viewed as saying that the function to be differentiated is differentiable on its domain and that the derivative is as given. For each, there is also a functional form,

EX: $(cf)' = cf'$, and a Liebniz form,

EX: $\dfrac{d}{dx}(cu) = c\dfrac{du}{dx}$

■ **Sum:**

$$\frac{d}{dx}[f(x) + g(x)] = f'(x) + g'(x)$$

■ **Scalar Multiple:**

$$\frac{d}{dx}[cf(x)] = cf'(x)$$

■ **Product:**

$$\frac{d}{dx}[f(x)\,g(x)] = f'(x)g(x) + f(x)g'(x)$$

■ **Quotient:**

$$\frac{d}{dx}\left[\frac{f(x)}{g(x)}\right] = \frac{f'(x)g(x) - f(x)g'(x)}{g(x)^2}$$

■ **The Chain Rule (for Compositions):**

$$\frac{d}{dx}[(f \circ g)(x)] \equiv \frac{d}{dx}f(g(x)) = f'(g(x))\,g'(x)$$

◆ This says that a small change in input to the composition is scaled by $g'(x)$, then by $f'(g(x))$.

◆ In Liebniz notation, if $z = f(y)$ and $y = g(x)$, and we thereby view z a function of x, then

$\dfrac{dz}{dx} = \dfrac{dz}{dy}\dfrac{dy}{dx}$, $\dfrac{dz}{dy}$ being evaluated at $y = g(x)$.

◆ In D notation, $D(f \circ g) = [D(f) \circ g]\, D(g)$.

■ **Inverse Functions:** If f is the inverse of a function g (and g' is continuous and nonzero), then

$$f'(x) = \frac{1}{g'(f(x))}.$$

◆ To get a specific formula directly, start with $y = f(x)$; rewrite it $g(y) = x$; differentiate with respect to x to get $g'(y)y' = 1$; write this $y' = 1/g'(y)$ and put $g'(y)$ in terms of x, using the relations $y = f(x)$ and $g(y) = x$.

EX: $y = \ln x$; $e^y = x$; $e^y y' = 1$; $y' = 1/e^y = 1/x$.

■ **Implicit Functions:** The derivative of a function defined implicitly by a relation $F(x,y) = c$ may be found by differentiating the relation with respect to x while treating y as a function of x wherever it appears in the relation; and then solving for y' in terms of x and y.

◆ The result is the same as obtained from the formal expression $y' = -\dfrac{\frac{d}{dx}F(x,y)}{\frac{d}{dy}F(x,y)}$, where y is treated as a constant in the numerator, x as a constant in the denominator.

Derivative Formulas

■ **Constants:** For any constant C, $\dfrac{d}{dx} C = 0$.

■ **Reciprocal Function:** $\dfrac{d}{dx} \dfrac{1}{x} = -\dfrac{1}{x^2}$

◆ The chain rule gives $\dfrac{d}{dx}\left[\dfrac{1}{f(x)}\right] = -\dfrac{1}{f(x)^2}\, f'(x)$.

■ **Square Root:** $\dfrac{d}{dx}\sqrt{x}=\dfrac{1}{2\sqrt{x}}$

■ **Powers:** For any real value of n, $\dfrac{d}{dx}x^n = nx^{n-1}$, valid where x^{n-1} is defined.

♦ The chain rule gives $\dfrac{d}{dx}[f(x)]^n = n[f(x)]^{n-1}f'(x)$.

■ **Exponentials:** An exponential function has derivative proportional to itself, the proportionality factor being the natural logarithm of the base:

$$\frac{d}{dx}e^x = e^x, \ \frac{d}{dx}a^x = (\ln a)\,a^x$$

♦ The chain rule gives $\dfrac{d}{dx}e^{f(x)} = e^{f(x)}f'(x)$.

■ **Logarithms:** $\dfrac{d}{dx}\ln|x| = \dfrac{1}{x}$, $\dfrac{d}{dx}\log_a|x| = \dfrac{1}{(\ln a)x}$

♦ Same rules hold without absolute value, but the domain is restricted to $(0, \infty)$.

♦ The chain rule gives $\dfrac{d}{dx}\ln|f(x)| = \dfrac{f'(x)}{f(x)}$.

■ **Hyperbolic Functions:**

$$\sinh'x = \cosh x$$
$$\cosh'x = \sinh x$$
$$\operatorname{arcsinh}'x = \frac{1}{\sqrt{1+x^2}}$$
$$\operatorname{arccosh}'x = \frac{1}{\sqrt{x^2-1}}$$

■ **Trig Functions:**

$$\sin'x = \cos x$$
$$\cos'x = -\sin x$$
$$\tan'x = \sec^2 x$$
$$\cot'x = -\csc^2 x$$

$$\sec' x = \sec x \tan x$$
$$\csc' x = -\csc x \cot x$$
$$\arcsin' x = \frac{1}{\sqrt{1-x^2}} = -\arccos' x$$
$$\arctan' x = \frac{1}{1+x^2} = -\text{arccot}' x$$

5 **Analysis**

Local Features of Functions

■ **Neighborhoods:** In the following, "near" a point means in an open interval containing the point.
   ◆ Such an open interval is often called a **neighborhood** of the point.
■ **Continuity:** If a function is differentiable at a point, then it is continuous there.
■ **Critical Points:** A point c is a critical point of f if f is defined near c and either $f'(c) = 0$ or $f'(c)$ does not exist.
■ **Local Extrema:** A **local minimum** point of f is a point c with $f(x) \geq f(c)$ for x near c. A **local maximum** point of f is a point c with $f(x) \leq f(c)$ for x near c.
   ◆ If c is a local extremum point, then it is a critical point (this follows from definitions).
   ◆ **Relative extrema** are the same as local extrema.
■ **First Derivative Test:** Suppose c is a critical point of f, and f is continuous at c.
   ◆ If $f'(x)$ changes sign from *negative to positive* as x increases through c, then c is a local *minimum* point.

◆ If $f'(x)$ changes sign from *positive to negative* as x increases through c, then c is a local *maximum* point.
◆ If $f'(x)$ keeps the same sign, then c is not an extremum point.

■ **Second Derivative Test:** Suppose f is differentiable near a critical point c.
◆ If $f''(c) > 0$, then c is a local minimum point.
◆ If $f''(c) < 0$, then c is a local maximum point.

■ **Inflection Points:** If the graph of f has a tangent line (possibly vertical) at c and $f''(x)$ changes sign as x increases through c, then c, or the graph point $(c, f(c))$, is called an inflection point.

EX: $x^{1/3}$ has a vertical tangent and inflection point at **(0,0)**. An inflection point for f is an extremum for f'; the tangent line is locally steepest at such a point. The only possible inflection points are where $f''(x) = 0$ or $f''(x)$ does not exist.

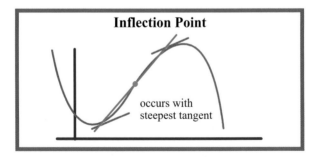

Inflection Point

occurs with steepest tangent

Trends & Global Features

■ **Mean Value Theorem** (MVT). If f is continuous on $[a, b]$ and differentiable on the open interval (a, b), then there is a point c in (a, b) with
$$f'(c) = \frac{f(b) - f(a)}{b - a}.$$

◆ Graphically, some tangent line between *a* and *b* is parallel to the secant line through *(a,f(a))* and *(b,f(b))*.

◆ The case with *f(a)* = *f(b)* = 0, whence *f'(c)* = 0, is **Rolle's Theorem**.

◆ The proof of the MVT relies on the **Extreme Value Theorem**.

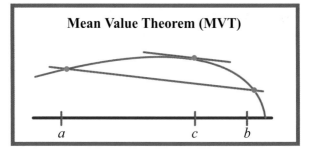

Mean Value Theorem (MVT)

■ **Increasing & Decreasing:**

◆ If *f'* = **0** on an interval, then *f* is constant on that interval.

◆ If *f'* > **0** on an interval, then *f* is strictly increasing on that interval.

◆ If *f'* < **0** on an interval, then *f* is strictly decreasing on that interval. (These follow from MVT.)

■ **Concavity:** A graph is said to be **concave up [down]** at a point *c* if the graph lies above [below] the tangent line near *c*, except at *c*.

◆ If *f''* > **0** on an interval, then the graph of *f* is concave up on the interval (UP-POSITIVE); also *f'* is increasing, and the tangent lines are turning upward as *x* increases.

◆ If *f''* < **0** on an interval, then the graph of *f* is concave down on the interval (DOWN-NEGATIVE); also *f'* is decreasing, and the tangent lines are turning downward as *x* increases.

■ **Extrema on a Closed Interval:** The global, or absolute, maximum and minimum values of a continuous function on a closed interval $[a,b]$ (guaranteed to be achieved by the Extreme Value Theorem) can only occur at critical points or endpoints.

Applications

■ **Optimization with Constraint:** Here is an outline to approach optimization problems involving two variables that are somehow related.

◆ **Visualize** the **problem** and name the variables.

◆ Write down the **objective function**—the one to be optimized—as a function of two variables.

◆ Write down a **constraint equation** relating the variables.

◆ Use the constraint to **rewrite** the **objective** function *in terms of one variable*.

◆ **Analyze** the new function of one variable to find its optimal point(s), and the optimal value.

EX: To maximize the area of a rectangle with perimeter being p, we pose the problem as maximizing $A = lw$ subject to the constraint $2l+2w = p$. The constraint gives $w = p/2 - l$, when $A = l(p/2-l)$. The maximum occurs at $l=p/4$, with $A = (p/4)^2$. A verbal result is clearest: it's a square. For geometric problems, **volume formulas** may be needed:

cylinder: πr^2h,
cone: $\pi r^2h/3$,
sphere: $4\pi r^3/3$.

■ **Cubics:** A cubic $p(x) = ax^3 + bx^2 + cx + d$ has exactly one inflection point: (h,k) where $h = -b/(3a)$ and $k = p(h)$. A **normal form** is
$$p(x) = a(x-h)^3 + m(x - h) + k$$
where $m = bh + c$ is the slope at the inflection point.

   ◆ If m and a have opposite signs, the horizontal line through the inflection point meets the graph at two points, each a distance $\sqrt{-m/a}$ from the inflection point, and local extrema occur at points $1\sqrt{3} \approx 0.6$ times that distance.

6

Integration: The Basics

NOTES
This chapter explains the basics of **integration** to **beginning calculus** students.

Interpretations

■ **Area Under a Curve:** The integral of a nonnegative function over an interval gives the area under the graph of the function.

■ **Average Value:** The average value of f over an interval $[a,b]$ may be defined by average value =

$$\frac{1}{b-a}\int_a^b f(x)\,dx$$

◆ Often a rough estimate of an integral can be made by estimating the average value (by inspection of the graph, for example) and multiplying it by the length of the interval.

■ **Accumulated Change:** The integral of a rate of change gives the total change in the original quantity over the time interval.

EX: if $v(t)=s'(t)$ represents velocity, then $v(t)\Delta t$ is the approximate displacement occurring in the time increment t to $t+\Delta t$. Adding the displacements for all the time increments gives the approximate change in position over the entire time interval. In the limit of small time increments, one gets the integral $\int_a^b v(t)\,dt = s(b) - s(a)$, which is the total displacement.

Fundamental Theorem of Calculus

■ **Antiderivatives:** An antiderivative of a function f is a function F whose derivative is f: $F'(x) = f(x)$ for all x in some domain. Any two antiderivatives of a function on an interval differ by a constant. (This follows from MVT.)

  EX: arctan x and $-\arctan(1/x)$ are both antiderivatives of $1/(1+x^2)$ for $x > 0$. (They differ by $\pi/2$.) An antiderivative is also called an **indefinite integral**, though the latter term often refers to the entire family of antiderivatives.

■ **The Fundamental Theorem:** There are two parts:

  ◆ **Evaluating Integrals:** If f is continuous on $[a,b]$, and F is any antiderivative of f on that interval, then $\int_a^b f(x)dx = F(x)\Big|_a^b \equiv F(b) - F(a)$.

  ◆ **Constructing Antiderivatives:** If f is continuous on $[a,b]$, then the function $G(x) = \int_a^x f(w)\,dw$ is an antiderivative of f on (a,b): $G'(x) = f(x)$. (The one-sided derivatives of G agree with f at the endpoints.)

■ **Differentiation of Integrals:** To differentiate a function such as $x \mapsto \int_a^{x^2} f(w)\,dw$, view it as a composition $G(x^2)$, with G as above.

  ◆ The chain rule gives $\dfrac{d}{dx}G(x^2) = G'(x^2) \cdot 2x = 2xf(x^2)$.

7 Integration: Advanced

> **NOTES**
> This chapter explains **integration** in greater detail, outlining complexities for **intermediate** and **advanced calculus** students.

Definitions

■ **Heuristics:** The definite integral captures the idea of adding the values of a function over a continuum.

■ **Riemann Sum:** A suitably weighted sum of values. A definite integral is the limiting value of such sums.

◆ A Riemann sum of a function f defined on $[a,b]$ is determined by a **partition**, which is a finite division of $[a,b]$ into subintervals, typically expressed by $a = x_0 < x_1 \cdots < x_n = b$; and a **sampling** of points, one point from each subinterval, say c_i from $[x_{i-1}, x_i]$.

◆ The associated Riemann sum is:

$$\sum_{i=1}^{n} f(c_i)(x_i - x_{i-1}).$$

◆ A **regular partition** has subintervals all the same length, $\Delta x = (b-a)/n$, $x_i = a + i\Delta x$. A partition's **norm** is its maximum subinterval length.

◆ A **left sum** takes the left endpoint $c_i = x_{i-1}$ of each subinterval, a **right sum**, the right endpoint.

◆ An **upper sum** (of a continuous f takes a point c_i in each subinterval where the maximum value of f is achieved, a **lower sum**, the minimum value.

EX: The upper Riemann sum of **cos$x$** on **[0,3]** with a regular partition of **n** intervals is the left sum (since the cosine is decreasing on the interval): $\sum_{i=1}^{n}\left[\cos\left((i-1)\frac{3}{n}\right)+1\right]\frac{3}{n}$.

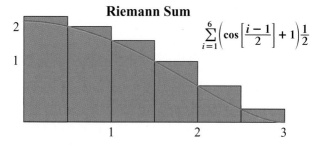

Riemann Sum

$$\sum_{i=1}^{6}\left(\cos\left[\frac{i-1}{2}\right]+1\right)\frac{1}{2}$$

■ **Definite Integral:** The **definite integral** of f from **a** to **b** may be described as

$$\int_{a}^{b}f(x)dx = \lim_{\|\Delta x\|\to 0}\sum_{i}f(c_i)\Delta S_i.$$

◆ The limit is said to exist if some number **S** (to be called the integral) satisfies the following: Every **$\varepsilon > 0$** admits a **δ** such that all Riemann sums on partitions of **[a,b]** with norm less than **δ** differ from **S** by less than **ε**. If there is such a value **S**, the function is said to be **integrable** and the value is denoted $\int_{a}^{b}f(x)dx$ or $\int_{a}^{b}f$. The function must be bounded to be integrable.

◆ The function f is called the **integrand** and the points a and b are called the **lower limit** and **upper limit** of integration, respectively. The word **integral** refers to the formation of $\int_a^b f$ from f and $[a,b]$ as well as to the resulting value if there is one.

■ **Antiderivative:** An antiderivative of a function f is a function A whose derivative is f: $A'(x)=f(x)$ for all x in some domain (usually an interval). Any two antiderivatives of a function on an interval differ by a constant (a consequence of the **Mean Value Theorem [MVT]**).

EX: both $\frac{1}{2}(x-a)^2$ and $\frac{1}{2}x^2-ax$ are antiderivatives of $x-a$, differing by $\frac{1}{2}a^2$. The **indefinite integral** of a function f, denoted $\int f(x)dx$, is an expression for the family of antiderivatives on a typical (often unspecified) interval.

EX: (for $x<-1$, or for $x>1$).
$$\int \frac{x}{\sqrt{x^2-1}} dx = \sqrt{x^2-1}+C.$$

◆ The constant C, which may have any real value, is the **constant of integration**. (Computer programs, and this chart, may omit the constant, it being understood by the knowledgeable user that the given antiderivative is just one representative of a family.)

Interpretations

■ **Area Under a Curve:** If f is nonnegative and continuous on $[a,b]$, then $\int_b^a f(x)dx$ gives the area between the x-axis and the graph.

◆ The **area function** $A(x)=\int_a^x f(t)dt$ gives the area accumulated up to x.

◆ If f is negative, the integral is the negative of the area.

■ **Average Value:** The average value of f over an interval $[a,b]$ may be defined by:

$$average\ value = \frac{1}{b-a}\int_a^b f(x)dx.$$

◆ A rough estimate of an integral may be made by estimating the average value (by inspecting the graph) and multiplying it by the length of the interval.

■ **Accumulated Change:** The integral of a rate of change of a quantity over a time interval gives the total change in the quantity over the time interval.

EX: if $v(t)=s'(t)$ is a velocity (the rate of change of position), then $v(t)\Delta t$ is the approximate displacement occurring in the time increment t to $t+\Delta t$; adding the displacements for all time increments gives the approximate change in position over the entire time interval. In the limit of small time increments, one gets the exact total displacement:

$$\int_a^b v(t)dt = s(b)-s(a)$$

■ **Integral Curve:** Imagine that a function f determines a slope $f(x)$ for each x. Placing line segments with slope $f(x)$ at points (x, y) for various y, and doing this for various x, one gets a **slope field**.

◆ An antiderivative of f is a function whose graph is tangent to the slope field at each point. The graph of the antiderivative is called an **integral curve** of the slope field.

■ **Solution to Initial Value Problem (IVP):** The solution to the differential equation $y'=f(x)$ with initial value $y(x_0)=y_0$ is $(x)=y_0+\int_{x_0}^{x} f(t)dt.$

Theory

■ **Integrability & Inequalities:** A continuous function on a closed interval is integrable. Integrability on $[a,b]$ implies integrability on closed subintervals of $[a,b]$.

◆ Assuming f is integrable, if $L \leq f(x) \leq M$ for all x in $[a,b]$, then $L \cdot (b-a) \leq \int_a^b f(x)dx \leq M \cdot (b-a)$.

◆ Use this to check integral evaluations with rough overestimates or underestimates.

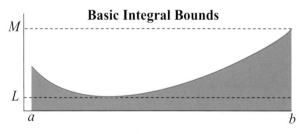

Basic Integral Bounds

◆ If f is nonnegative, then $\int_a^b f(x)dx$ is nonnegative.

◆ If f is integrable on $[a,b]$, then $\left| \int_a^b f(x)dx \right| \leq \int_a^b |f(x)|dx$.

■ Fundamental Theorem of Calculus:

◆ One part of the theorem is used to **evaluate integrals**: If f is continuous on $[a,b]$, and A is an antiderivative of f on that interval, then

$$\int_a^b f(x)dx = A(x)\Big|_a^b \equiv A(b) - A(a).$$

◆ The other part is used to **construct antiderivatives:** If f is continuous on $[a,b]$, then the function $A(x) = \int_a^x f(t)dt$ is an antiderivative of f on $[a,b]$: $A'(x) = f(x)$ (valid for one-sided derivatives at the endpoints).

Fundamental Theorem

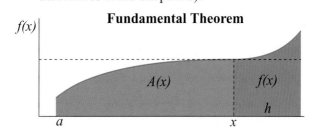

■ Differentiation of Integrals: Functions are often defined as integrals.

EX: the "sine integral function" is

$$Si(x) = \int_0^x \left(\frac{\sin t}{t}\right)dt.$$

◆ To differentiate such, use the second part of the fundamental theorem: $Si'(x) = \sin x / x$.

◆ A function such as $\int_a^{x^2} f(t)dt$ is a composition involving $A(u) = \int_a^u f(t)dt$.

◆ To differentiate, use the chain rule and the fundamental theorem:

$$\frac{d}{dx}\int_a^{x^2} f(t)dt = \frac{d}{dx}A(x^2) = A'(x^2)2x = 2xf(x^2).$$

■ **Mean Value Theorem (MVT) for Integrals:** If f and g are continuous on $[a,b]$, then there is a ξ in $[a,b]$ such that $\int_a^b f(x)g(x)dx = f(\xi)\int_a^b g(x)dx$.

◆ In the case $g \equiv 1$, the average value of f is attained somewhere on the interval:
$$\frac{1}{b-a}\int_a^b f(x)dx = f(\xi).$$

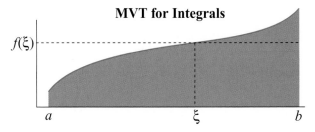

MVT for Integrals

■ **Change of Variable Formula:** An integrand and limits of integration can be changed to make an integral easier to apprehend or evaluate. In effect, the "area" is smoothly redistributed without changing the integral's value.

◆ If g is a function with continuous derivative and f is continuous, then $\int_a^b f(u)du = \int_c^d f(g(t))g'(t)dt$, where c,d are points with $g(c) = a$ and $g(d) = b$.

◆ In practice, **substitute** $u = g(t)$; compute $du = g'(t)dt$; and find what t is when $u = a$ and $u = b$.

EX: $u = \sin t$ effects the transformation

$$\int_a^b \sqrt{1-u^2}\, du = \int_0^{\pi/2}\sqrt{1-\sin^2 t}\, \cos t\, dt, \text{ which}$$

becomes $\int_0^{\pi/2} \cos^2 t\, dt$, since $\sqrt{1-\sin^2 t} = \cos t$ for $0 \le t \le \pi/2$. The formula is often used in reverse, starting with $\int_b^a F(g(x))g'(x)\,dx$.

■ **Natural Logarithm:** A rigorous definition is $\ln x = \int_1^x \frac{1}{u}\,du$.

◆ The change of variable formula with $u = 1/t$ yields $\int_1^{1/x} \frac{1}{u}\,du = \int_1^x t\frac{-1}{t^2}\,dt = -\int_1^x \frac{1}{t}\,dt$, showing that $\ln(1/x) = -\ln x$.

◆ The other elementary properties of the natural log can likewise be easily derived from this definition. In this approach, an inverse function is deduced and is defined to be the natural exponential function.

Integration Formulas

■ **Basic Indefinite Integrals:** Each formula gives just one antiderivative (all others differing by a constant from that given), and is valid on any open interval where the integrand is defined.

$$\int x^n\,dx = \frac{x^{n+1}}{n+1}(n \ne -1) \qquad \int \frac{1}{x}\,dx = \ln|x|$$

$$\int e^{kx}\,dx = \frac{e^{kx}}{k}(k \ne 0) \qquad \int a^x\,dx = \frac{a^n}{\ln a}(a \ne 1)$$

$$\int \cos x\,dx = \sin x \qquad \int \sin x\,dx = -\cos x$$

$$\int \frac{dx}{1+x^2} = \arctan x \qquad \int \frac{dx}{\sqrt{1-x^2}} = \arcsin x$$

■ **Further Indefinite Integrals:** The above conventions hold.

$$\int \tan x \, dx = \ln|\sec x| \qquad \int \cot x \, dx = \ln|\sin x|$$

$$\int \sec x \, dx = \ln|\sec x + \tan x|$$

$$\int \csc x \, dx = \ln|\csc x + \cot x|$$

$$\int \cosh x \, dx = \sinh x \qquad \int \sinh x \, dx = \cosh x$$

$$\int \frac{dx}{x^2 - a^2} = \frac{1}{2a}\ln\left|\frac{x-a}{x+a}\right| \qquad \int |x| \, dx = \frac{1}{2}x|x|$$

$$\int \frac{dx}{\sqrt{x^2+a^2}} = \ln\left|x+\sqrt{x^2+a^2}\right| = \sinh^{-1}\frac{x}{a} + \ln a$$

$$\int \frac{dx}{\sqrt{x^2-a^2}} = \ln\left|x+\sqrt{x^2-a^2}\right| = \cosh^{-1}\frac{x}{a} + \ln a$$

(take positive values for **cosh⁻¹**)

$$\int \sqrt{x^2 \pm a^2} \, dx = \frac{1}{2}x\sqrt{x^2 \pm a^2} \pm \frac{a^2}{2}\ln\left|x+\sqrt{x^2 \pm a^2}\right|$$

(Take same sign, + or –, throughout)

$$\int \sqrt{a^2-x^2} \, dx = \frac{1}{2}x\sqrt{a^2-x^2} + \frac{a^2}{2}\arcsin\frac{x}{a}$$

■ **Common Definite Integrals:**

$$\int_0^1 x^n \, dx = \frac{1}{n+1} \qquad \int_0^r \sqrt{r^2-x^2} \, dx = \frac{\pi r^2}{4}$$

$$\int_0^\pi \sin x \, dx = 2$$

$$\int_0^{\pi/2} \cos^2\theta \, d\theta = \int_0^{\pi/2} \frac{1+\cos 2\theta}{2} \, d\theta = \frac{\pi}{4}$$

$$\int_0^{\pi/2} \sin^2\theta \, d\theta = \int_0^{\pi/2} \frac{1-\cos 2\theta}{2} \, d\theta = \frac{\pi}{4}$$

◆ To remember which of $\frac{1}{2}(1\pm\cos\ 2\theta)$ equals $\cos^2\theta$ or $\sin^2\theta$, recall the value at zero.

Techniques

■ **Substitution:** Refers to the **Change of variable formula,** but often the formula is used in reverse.

◆ For an integral recognized to have the form $\int_a^b F(g(x))g'(x)dx$ (with F and g' continuous), you can put $u=g(x)$, $du=g'(x)dx$, and modify the limits of integration appropriately:
$$\int_a^b F(g(x))g'(x)dx = \int_{g(a)}^{g(b)} F(u)du.$$

◆ In effect, the integral is over a path on the u-axis traced out by the function g. (If $g(b) = g(a)$ [the path returns to its start], then the integral is zero.)

EX: $u=1+x^2$ yields
$$\int_0^1 \frac{x}{1+x^2}dx=\frac{1}{2}\int_0^1 \frac{1}{1+x^2}2xdx=\frac{1}{2}\ln(1+x^2).$$

◆ Substitution may be used for indefinite integrals.
$$\int \frac{x}{1+x^2}dx=\frac{1}{2}\int \frac{du}{u}=\frac{1}{2}\ln u=\frac{1}{2}\ln(1+x^2).$$

Some **general formulas** are:
$$\int g(x)^n g'(x)dx=\frac{g(x)^{n+1}}{n+1},$$
$$\int \frac{g'(x)}{g(x)}dx=\ln|g(x)|,$$
$$\int e^{g(x)}g'(x)dx=e^{g(x)}.$$

■ **Integration by Parts:** Explicitly,

$$\int_a^b u(x)v'(x)dx = u(x)v(x)\Big|_a^b - \int_a^b v(x)u'(x)dx.$$

The common formula is $\int_a^b u\,dv = uv\Big|_a^b - \int_a^b v\,du$.

For indefinite integration, $\int u\,dv = uv - \int v\,du$.

◆ The procedure is used in derivations where the functions are general, as well as in explicit integrations. You don't need to use "u" and "v."

◆ View the integrand as a product with one factor to be integrated and the other to be differentiated; the integral is the integrated factor times the one to be differentiated, minus the integral of the product of the two new quantities.

◆ The factor to be integrated may be **1** (giving $v=x$).

EX: $\int \arctan x\,dx = x \arctan x - \int \dfrac{x}{1+x^2}dx$

Other routine integration-by-parts integrands are $\arcsin x$, $\ln x$, $x^n \ln x$, $x \sin x$, $x \cos x$, and xe^{ax}.

■ **Rational Functions:** Every rational function may be written as a polynomial plus a proper rational function (degree of numerator less than degree of denominator).

◆ A proper rational function with real coefficients has a **partial fraction decomposition**: it can be written as a sum with each summand being either a constant over a power of a linear polynomial or a linear polynomial over a power of a quadratic.

◆ A factor $(x-c)^k$ in the denominator of the rational function implies there could be summands $\dfrac{A_1}{x-c} + \ldots + \dfrac{A_k}{(x-c)^k}$.

◆ A factor $(x^2+bx+c)^k$ (the quadratic not having real roots) in the denominator implies there could be summands
$$\frac{A_1+B_1x}{x^2+bx+c}+\dots+\frac{A_k+B_kx}{(x^2+bx+c)^k}.$$

◆ Math software can handle the work, but the following case should be familiar. If $a\neq b$,
$$\frac{1}{(x-a)(x-b)}=\frac{C}{x-a}+\frac{D}{x-b}$$ where C, D are seen to be $C=-D=\frac{1}{a-b}$. Thus
$$\int\frac{1}{(x-a)(x-b)}dx=\frac{1}{a-b}\left(\ln\left|x-a\right|-\ln\left|x-b\right|\right).$$

◆ In general, the indefinite integral of a proper rational function can be broken down via partial fraction decomposition and linear substitutions (of form $u=ax+b$) into the integrals $\int u^{-1}du$, $\int u^{-n}du (n>1)$, $\int u(u^2+1)^{-n}du$ (handled with substitution $w=u^2+1$), and $\int(u^2+1)^{-n}du$ (handled with substitution $u=\tan t$).

Improper Integrals

■ **Unbounded Limits:** If f is defined on $[a,\infty]$ and integrable on $[a,B]$ for all $B>a$, then $\int_a^\infty f(x)dx\overset{\text{def}}{=}\lim_{B\to\infty}\int_a^B f(x)dx$ provided the limit exists.

EX: $\int_0^\infty e^{-x}dx=\lim_{B\to\infty}(1-e^{-B})=1$

Likewise, for appropriate f,

$$\int_{-\infty}^{b} f(x)dx \overset{def}{=} \lim_{A \to -\infty} \int_{A}^{b} f(x)dx.$$

◆ In each case, if the limit exists, the improper integral **converges**; otherwise it **diverges**. For f defined on $(-\infty, \infty)$ and integrable on every bounded interval,

$$\int_{-\infty}^{\infty} f(x)dx \overset{def}{=} \lim_{A \to -\infty} \int_{A}^{c} f(x)dx + \lim_{B \to \infty} \int_{c}^{B} f(x)dx$$

(the choice of c being arbitrary), provided each integral on the right converges.

■ **Singular Integrands:** If f is defined on $(a,b]$ but not at $x=a$ and is integrable on closed subintervals of $(a,b]$, then $\int_{a}^{b} f(x)dx \overset{def}{=} \lim_{c \to a^{+}} \int_{c}^{b} f(x)dx$ provided the limit exists.

◆ A similar definition holds if the integrand is defined on $[a,b)$.

EX: $\int_{0}^{2} \dfrac{1}{\sqrt{4-x^2}} dx$ is $\lim\limits_{c \to 2^{-}} \int_{0}^{c} \dfrac{1}{\sqrt{4-x^2}} dx =$

$\lim\limits_{c \to 2^{-}} \arcsin\left(\dfrac{c}{2}\right) = \dfrac{\pi}{2}.$

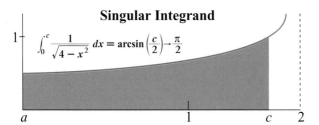

Singular Integrand

$\int_{0}^{c} \dfrac{1}{\sqrt{4-x^2}} dx = \arcsin\left(\dfrac{c}{2}\right) \to \dfrac{\pi}{2}$

◆ If f is not defined at a finite number of points in an interval $[a,b]$, and is integrable on closed subintervals of open intervals between such points, the integral $\int_a^b f$ is defined as a sum of left and right-hand limits of integrals over appropriate closed subintervals, provided all the limits exist.

EX: $\int_{-1}^{1} \frac{1}{x^3} dx = \lim_{a \to 0^-} \int_{-1}^{a} \frac{1}{x^3} dx + \lim_{b \to 0^+} \int_{b}^{1} \frac{1}{x^3} dx$

if the limits on the right were to exist. They don't, so the integral diverges.

■ **Examples & Bounds:**

$\int_1^{\infty} \frac{1}{x^p} dx$ converges for $p > 1$; diverges otherwise.

$\int_0^1 \frac{1}{x^p} dx$ converges for $p < 1$; diverges otherwise.

$\int_2^{\infty} \frac{1}{x(\ln x)^p} dx$ converges for $p > 1$; diverges otherwise.

Note: $\int \frac{dx}{x(\ln x)^p} = -\frac{1}{(n-1)(\ln x)^{p-1}}$,

$p > 1$ converges at $x = \infty$, $p = 0$ or < 1 diverges at $x = \infty$.

EX: $\int_1^{\infty} \frac{1}{x^2} dx$ converges to 1; and $\int_0^1 \frac{1}{x^2} dx$

diverges. The above integrals are useful in comparisons to establish convergence (or divergence) and to get bounds.

EX: $\int_0^{\infty} \left(\frac{x}{1+x^2} \right)^{3/2} dx$ converges since the integrand

is bounded by $1/2^{3/2}$ on $[0, 1]$ and is always less than $1/x^{3/2}$. It converges to a number less than

$$\frac{1}{2^{3/2}} \int_0^{\infty} \left(\frac{x}{1+x^2} \right)^{3/2} dx = \frac{1}{2^{3/2}} + 2 < 2.4.$$

Applications

8

NOTES
This chapter explains the **application** of calculus principles in other branches of mathematics and in science. [For **intermediate** and **advanced calculus** students.]

Geometry

■ **Areas of Plane Regions:** Consider a plane region admitting an axis such that sections perpendicular to the axis vary in length according to a known function $L(p)$, $a \leq p \leq b$.

◆ The area of a strip of width Δp perpendicular to the axis at p is $\Delta A = L(p)\Delta p$, and the total area is

$$A = \int_a^b L(p)\,dp.$$

EX: the area of the region bounded by the graphs of f and g over $[a,b]$ is $\int_a^b \left[g(x) - f(x)\right] dx$ provided $g(x) \geq f(x)$ on $[a,b]$. Sometimes it is simpler to view a region as bounded by two graphs "over" the y-axis, in which case the integration variable is y.

Planar Area

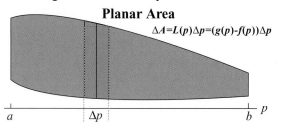

$\Delta A = L(p)\Delta p = (g(p) - f(p))\Delta p$

■ **Volumes of Solids:**
Consider a solid
admitting an axis
such that cross-
sections perpendi-
cular to the axis vary
in area according to
a known function
$A(p)$, $a \leq p \leq b$.

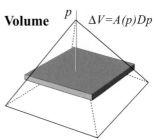

◆ The volume of a slab of thickness Δp perpendicular to the axis at p is $\Delta V = A(p) \Delta p$, and the
total volume is $V = \int_a^b A(p) dp$.

EX: a pyramid having square horizontal cross-
sections, with bottom side length s and height
h, has cross-sectional area $A(z) = [s(1 - z/h)]^2$
at height z. Its volume is thus
$$V = \int_0^h s^2 \left(1 - \frac{z}{h}\right)^2 dz = \frac{s^2 h}{3}.$$

■ **Solids of Revolution:** Consider a solid of
revolution determined by a known radius function
$r(z)$, $a \leq z \leq b$, along its axis of revolution.

◆ The area of the cross-sectional "**disk**" at z is
$A(z) = \pi r(z)^2$, and the volume is
$$V = \int_a^b A(z) dz = \int_a^b \pi r(z)^2 dz.$$

◆ If the solid lies between two radii $r_1(z)$ and $r_2(z)$ at
each point z along the axis of revolution, the
cross-sections are "**washers**," and the volume is
the obvious difference of volumes like that above.

◆ Sometimes a radial coordinate r, $a \leq r \leq b$, along
an axis perpendicular to the axis of revolution
parametrizes the heights $h(r)$ of cylindrical

sections (**shells**) of the solid parallel to the axis of revolution.

◆ In this case, the area of the shell at r is $A(r)=2\pi r h(r)$, and the volume of the solid is

$$V=\int_a^b A(r)dr=\int_a^b 2\pi rh(r)dr.$$

■ **Arc Length:**

◆ A graph $y=f(x)$ between $x=a$ and $x=b$ has length $V=\int_a^b \sqrt{1+f'(x)^2}\,dx$.

◆ A curve C parametrized by $(x(t), y(t))$, $a\leq t\leq b$, has length $\int_C ds=\int_a^b \sqrt{x'(t)^2+y'd(t)^2}\,dt$.

■ **Area of a Surface of Revolution:** The surface generated by revolving a graph $y=f(x)$ between $x=a$ and $x=b$ about the x-axis has area $\int_b^a 2\pi f(x)\sqrt{1+f'(x)^2}\,dx$.

◆ If the generating curve C is parametrized by $(x(t), y(t))$, $a\leq t\leq b$, and is revolved about the x axis, the area is

$$\int_C 2\pi yds=\int_a^b 2\pi y(t)\sqrt{x'(t)^2+y'(t)^2}\,dt.$$

Physics

■ **Motion in One Dimension:** Suppose a variable displacement $x(t)$ along a line has velocity $v(t)=x'(t)$ and acceleration $a(t)=v'(t)$. Since v is an antiderivative of a, the fundamental theorem implies:

$$v(t) = v(t_0) + \int_{t_0}^t a(u)du, \; x(t) = x(t_0) + \int_{t_0}^t v(u)du.$$

EX: the height $x(t)$ of an object thrown at time $t_0=0$ from a height $x(0)=x_0$ with a vertical velocity $v(0)=v_0$ undergoes the acceleration $-g$ due to gravity. Thus $v(t)=v(v_0)+\int_0^t (-u)\,du=v_0-gt$ and $x(t)=x_0+\int_0^t (v_0-gu)\,du=x_0+v_0t-\frac{1}{2}gt^2$.

■ **Work:** If $F(x)$ is a variable force acting along a line parametrized by x, the approximate work done over a small displacement Δx at x is $\Delta W=F(x)\Delta x$ (force times displacement), and the work done over an interval $[a,b]$ is $W=\int_a^b F(x)\,dx$.

◆ In a **fluid lifting** problem, often $\Delta W=\Delta F \bullet h(y)$, where $h(y)$ is the lifting height for the "**slab**" of fluid at y with cross-sectional area $A(y)$ and width Δy, and the slab's weight is $\Delta F=\rho A(y)\Delta y$, ρ being the fluid's weight-density. Then $W=\int_a^b \rho A(y)h(y)\,dy$.

Differential Equations (DE)

■ A differential equation (DE) was solved in the item **solution to initial value problem**; an example of that type is in **motion in one dimension**. In those, the expression for the derivative involved only the independent variable.

◆ A basic DE involving the dependent variable is $y'=ky$.

◆ A general DE where only the **first-order** derivative appears and is **linear** in the dependent variable is $y'+p(t)y=q(t)$.

◆ Generally more difficult are equations in which the independent variable appears in a \hlt{nonlinear} way:

EX: $y' = y^2 - x$. Common in applications are **second-order** DEs that are linear in the dependent variable;

EX: $y'' = -ky$, $x^2 y'' + xy' + x^2 y = 0$.

■ **Solutions:** A solution of a DE on an interval is a function that is differentiable to the order of the DE and satisfies the equation on the interval.

◆ It is a **general solution** if it describes virtually all solutions, if not all.

◆ A general solution to an *n*th order DE generally involves *n* constants, each admitting a range of real values.

◆ An **initial value problem** (IVP) for an *n*th order DE includes a specification of the solution's value and *n*−1 derivatives at some point.

◆ Generally in applications, an IVP has a **unique solution** on some interval containing the initial value point.

■ **Basic First-Order Linear DE:** The equation $y' = ky$ rewritten $\dfrac{dy}{dt} = ky$ suggests $\dfrac{dy}{y} = k\,dt$ where

$\left| y \right| = kt + c$. In this way one finds a solution $y = Ce^{kt}$.

◆ On any open interval, every solution must have that form, because $y' = ky$ implies $\dfrac{d}{dt}(ye^{-kt})$, where ye^{-kt} is constant on the interval. Thus $y = Ce^{kt}$ (*C* real) is the general solution.

◆ The unique solution with $y(a)=y_a$ is $y=y_a e^{k(t-a)}$.

◆ The **trivial solution** is $y\equiv0$, solving any IVP $y(a)=0$.

■ **General First-Order Linear DE:** Consider $y'+p(t)y=q(t)$. The solution to the associated **homogeneous equation $h'+p(t)h=0$** $(dh/h=-p(t)dt)$ with $h(a)=1$ is $h(t)=exp\left[-\int_a^t p(u)du\right]$.

◆ If y is a solution to the original DE, then $(y/h)'=q/h$, where $y=h\int q/h$.

◆ The solution with $y(a)=y_a$ is
$$y(t)=y_a+\left[-\int_a^t q(u)h(u)^{-1}du\right].$$

Approximations

Taylor's Formula

■ **Taylor Polynomials:** The **nth degree** Taylor polynomial of f at c is $P_n(x)=f(c)+f'(c)(x-c)+$

$\frac{1}{2!}f''(c)(x-c)^2+...+\frac{1}{n!}f^{(n)}(c)(x-c)^n$ (provided the derivatives exist).

◆ When $c=0$, it's also called a **MacLaurin polynomial**.

■ **Using Taylor's Formula:** Assume f has $n+1$ continuous derivatives on open interval and that c is a point in the interval. Then for any x in the interval, $f(x)=P_n(x)+R_n(x)$, where

$$R_n(x)=\frac{1}{(n+1)!}f^{(n+1)}(\xi)\cdot(x-c)^{n+1} \quad \text{for some} \quad \xi$$

between c and x (ξ varying with x).

◆ The expression for $R_n(x)$ is called the **Lagrange form of the remainder**.

　EX: The remainders for the MacLaurin polynomials of $f(x)=\ln(1+x)$, $-1<x<1$, are

$$R_n(x)=\frac{(-1)^n}{(n+1)(1+\xi)^{n+1}}\cdot x^{n+1}.$$

There is a ξ between 0 and x such that
$$\ln(1+x) = x - \frac{x^2}{2} + \frac{1}{3(1+\xi)^3}.$$

■ **Error Bounds:** As x approaches c, the remainder generally becomes smaller, and a given Taylor polynomial provides a better approximation of the function value.

◆ With the assumptions and notation above, if $\left| f^{(n+1)}(x) \right|$ is bounded by M on the interval, then $\left| f(x) - P_n(x) \right| \leq \frac{M}{(n+1)!} \left| x-c \right|^{n+1}$ for all x in the interval.

EX: For $|x| < 1$, $e^x \approx 1 + x + x^2/2$, with error no more than $\frac{3}{3!} |x|^3 = 0.5 |x|^3$, because the third derivative of e^x is bounded by 3 on $(-1,1)$.

■ **Big O Notation:** The statement $f(x) = p(x) + O(x^m)$ (as $x \to 0$) means that $\frac{f(x) - p(x)}{x^m}$ is bounded near $x=0$. (Some authors require that the limit of this ratio as x approaches 0 exist.) That is, $f(x) - p(x)$ approaches 0 at essentially the same rate as x^m

EX: Taylor's formula implies $f(x) = f(0) \neq f'(0)x + \frac{1}{2}f''(0)x^2 + O(x^3)$ if f has continuous third derivative on an open interval containing 0.

EX: $\sin x = x + O(x^3)$.

[Similar relations can be inferred from the identities in **Basic MacLaurin Series**, chapter 10, page 80.]

■ **L'Hôpital's Rule:** This resolves indeterminate ratios or $\left(\dfrac{0}{0} \text{ or } \dfrac{\infty}{\infty}\right)$. If $\lim\limits_{x \to a} f(x) = 0 = \lim\limits_{x \to a} g(x)$ and if $\lim\limits_{x \to a} f'(x) = 0 = \lim\limits_{x \to a} g'(x)$ are defined and $g(x) \neq 0$ for x near a (but not necessarily at a), then $\lim\limits_{x \to a} \dfrac{f(x)}{g(x)} = 0 = \lim\limits_{x \to a} \dfrac{f'(x)}{g'(x)}$ provided the latter limit exists, or is infinite.

◆ The rule also holds when the limits of f and g are infinite. (Note that $f'(a)$ and $g'(a)$ are not required to exist.)

◆ To resolve an indeterminate difference ($\infty - \infty$), try to rewrite it as an indeterminate ratio and apply L'Hôpital's rule.

◆ To resolve an indeterminate exponential ($0^0, 1^\infty, \text{or } \infty^0$), take its logarithm to get a product and rewrite this as a suitable indeterminate ratio. Apply L'Hôpital's rule: the exponential of the result resolves the original indeterminate exponential.

For $\lim\limits_{x \to 0} |x|^x$ you get and find $\lim\limits_{x \to 0} \dfrac{\ln|x|}{1/x} = \lim\limits_{x \to 0} \dfrac{1/x}{-1/x^2} = 0$, where $\lim\limits_{x \to 0} |x|^x = e^0 = 1$.

Numerical Integration

■ Solutions to applied problems often involve definite integrals that cannot be evaluated easily, if at all, by finding antiderivatives. Readily available software using refined algorithms can evaluate many integrals to needed precision.

◆ The following methods for approximating $\int_b^a f(x)\,dx$ are elementary. Throughout, n is the number of intervals in the underlying regular partition and $h=(b-a)/n$.

■ **Trapezoid Rule:** The line connecting two points on the graph of a positive function together with the underlying interval on the x axis form a trapezoid whose area is the average of the two function values times the length of the interval. Adding these areas up over a regular partition gives the trapezoid rule approximation

$$T_n=\left(\frac{f(a)}{2}+\sum_{i=1}^{n-1} f(a-ih)+\frac{f(b)}{2}\right)h.$$

◆ This is also the average of the left sum and right sum for the given partition. The approximation remains valid if f is not positive.

■ **Midpoint Rule:** This evaluates the Riemann sum on a regular partition with the sampling given by the midpoints of each interval:

$$M_n=\sum_{I=0}^{n-1} f\left(a+\left[i-\frac{1}{2}\right]h\right)h$$

◆ Each summand is the area of a trapezoid whose top is the tangent line segment through the midpoint.

■ **Simpson's Rule:** The weighted sum $\frac{1}{3}T_1+\frac{2}{3}M_1$ on the interval $[a,b]$ yields Simpson's rule

$$S = \frac{b-a}{6}\left(f(a)+4f\left(\frac{a+b}{2}\right)+f(b)\right).$$

◆ This is also the integral of the quadratic that interpolates the function at the three points. For a regular partition of $[a,b]$ into an even number

$n=2m$ of intervals, a formula is: $S_{2m} = \frac{h}{3}\{f(a) +$

$4 \sum_{i=0}^{m-1} f(a) + [2i + 1]h + 2 \sum_{i=0}^{m-1} f(a + 2i \cdot h) +$

$f(b)\}$ where $h = (b - a)/n$.

◆ Simpson's rule is exact on cubics.

Simpson's Rule

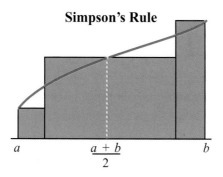

10

Sequences & Series

> **NOTES**
> This chapter explains **sequences** and **series** to **intermediate** and **advanced calculus** students.

Sequences

- **Sequences:** Functions whose domains consist of all integers greater than or equal to some initial integer, usually **0** or **1**, are called **sequences**.
  - ◆ The integer in a sequence at *n* is usually denoted with a subscripted symbol like a_n (rather than with a functional notation $a(n)$) and is called a **term** of the sequence.
  - ◆ A sequence is often referred to with an expression for its terms.
  - **EX:** $1/n$ (with the domain understood), in lieu of a fuller notation like: $\{1/n\}_{n=1}^{\infty}$, or $n \mapsto 1/n \, (n=1,2,\ldots)$.

- **Elementary Sequences:** An **arithmetic sequence**, with terms a_n, has a common difference *d* between successive terms: $a_n = a_{n-1} + d = a_0 + d \cdot n$.
  - ◆ It is a sequential version of a linear function, the common difference in the role of slope.
  - ◆ A **geometric sequence, with terms** g_n, has a common ratio *r* between successive terms: $g_n = g_{n-1}r = g_0 r^n$.

EX: 5.0, **2.5**, **1.25**, **0.625**, **0.3125,...** It is a sequential version of an exponential function, the common ratio in the role of base.

■ **Convergence:** A sequence $\{a_n\}$ **converges** if some number L (called the limit) satisfies the following: Every $\varepsilon > 0$ admits an N such that $|a_n - L| < \varepsilon$ for all $n \geq N$. If a limit L exists, there is only one; one says $\{a_n\}$ converges to L, and writes $a_n \to L$, or $\lim\limits_{n \to \infty} a_n = L$.

■ If a sequence does not converge, it **diverges**. If a sequence $\{a_n\}$ diverges in such a way that every $M > 0$ admits an N such that $a_n > M$ for all $n \geq N$, then one writes $a_n \to \infty$.

EX: If $|r| < 1$ then $r^n \to 0$; if $r = 1$ then $r^n \to 1$; otherwise r_n diverges, and if $r > 1$, $r^n \to \infty$.

■ **Bounded Monotone Sequences:** An increasing sequence that is bounded above converges (to a limit less than or equal to any bound).
   ◆ This is a fundamental fact about the real numbers, and is basic to series convergence tests.

Series of Real Numbers

■ **Series:** A series is a sequence obtained by adding the values of another sequence $\sum\limits_{n=0}^{N} a_n = a_0 + ... + a_N$.

   ◆ The value of the series at N is the sum of values up to a_N and is called a **partial sum**: $\sum\limits_{n=0}^{N} a_n = a_0 + ... + a_N$.

◆ The series itself is denoted $\sum\limits_{n=0}^{\infty} a_n$.

◆ The a_n are called the **terms** of the series.

■ **Convergence:** A series $\sum\limits_{n=0}^{\infty} a_n$ **converges** if the sequence of partial sums converges, in which case the limit of the sequence of partial sums is called the **sum** of the series.

◆ If the series converges, the notation for the series itself stands also for its sum:

$$\sum_{n=0}^{\infty} a_n = \lim_{N \to \infty} \sum_{n=0}^{N} a_n.$$

◆ An equation such as $\sum\limits_{n=0}^{\infty} a_n = S$ means the series converges and its sum is S.

◆ In general statements, $\sum a_n$ may stand for $\sum\limits_{n=0}^{\infty} a_n = S$.

■ **Geometric Series:** A (numerical) geometric series has the form $\sum\limits_{n=0}^{\infty} ar^n$, where r is a real number and $a \neq 0$.

◆ A key identity is $\sum\limits_{n=0}^{N} r^n = 1 + r + r^2 + ... + r^N = \dfrac{1-r^{N+1}}{1-r}(r \neq 1)$. It implies $\sum\limits_{n=0}^{\infty} r^n = \dfrac{1}{1-r}\left(\text{if} |r| < 1\right)$

$\left(\text{also} \sum\limits_{n=1}^{\infty} ar^n = a\left(\dfrac{1}{1-r}-1\right)\right)$, and that the series diverges if $|r| > 1$.

◆ The series diverges if $r = \pm 1$. The convergence and possible sum of any geometric series can be determined using the preceding formula.

EX: $\sum\limits_{n=1}^{\infty} \dfrac{4}{3^n} = 4\left(\dfrac{1}{1 - \frac{1}{3}} - 1\right) = 2.$

■ **P-Series:** For p, a real number, $\sum\limits_{n=1}^{\infty} \dfrac{1}{n^p}$ is called the p-series.

◆ The p-series diverges if $p \leq 1$ and converges if $p > 1$ (by comparison with harmonic series and the integral test, below).

◆ The **harmonic series** $\sum\limits_{n=1}^{\infty} \dfrac{1}{n}$ diverges, for the partial sums are unbounded: $\sum\limits_{n=1}^{2^N} \dfrac{1}{n} \geq 1 + \dfrac{N}{2}.$

■ **Alternating Series:** These are series whose terms alternate in (nonzero) sign.

◆ If the terms of an alternating series strictly decrease in absolute value and approach a limit of zero, then the series converges.

◆ Moreover, the truncation error is less than the absolute value of the first omitted term:

$$\left| \sum_{n=1}^{\infty} (-1)^n a_n - \sum_{n=1}^{N} (-1)^n a_n \right| < a_{N+1}. \quad \text{(assuming}$$

$a_n \to 0$ in a strictly decreasing manner).

Convergence Tests

■ **Basic Considerations:** For any K, if $\sum_{n=K}^{\infty} a^n$ converges, then $\sum_{n=1}^{\infty} a^n$ converges; conversely, if $a_n \not\to 0$, then $\sum a_n$ diverges. (Equivalently, if $\sum a_n$ converges, then $a_n \to 0$).

◆ This says nothing about the following: $\sum_{n=K}^{\infty} \frac{1}{n}$.

◆ A series of positive terms is an increasing sequence of partial sums; if the sequence of partial sums is bounded, the series converges. This is the foundation of all the following criteria for convergence.

■ **Integral Test & Estimate:** Assume f is continuous, positive, and decreasing on (K, ∞). Then $\sum_{n=K}^{\infty} f(n)$ converges if and only if $\int_{K}^{\infty} f(x)dx$, converges.

◆ If the series converges, then $\sum_{n=K}^{\infty} f(n) \le \sum_{n=K}^{N} f(n) + \int_{N}^{\infty} f(x)dx$, the right side overestimating the sum with error less than $\sum_{n=1}^{\infty} \frac{1}{n^3} \approx$ $\sum_{n=1}^{12} \frac{1}{n^3} + \int_{13}^{\infty} \frac{1}{x^3}dx = 1.2018\cdots$, the left side underestimating the sum with error less than $f(N+1)$.

Integral Test

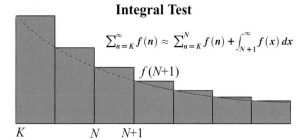

$$\sum_{n=K}^{\infty} f(n) \approx \sum_{n=K}^{N} f(n) + \int_{N+1}^{\infty} f(x)\, dx$$

$f(N+1)$

$K \qquad N \quad N+1$

EX: $\sum_{n=1}^{\infty} \dfrac{1}{n^3} \approx \sum_{n=1}^{12} \dfrac{1}{n^3} + \int_{13}^{\infty} \dfrac{1}{x^3}\, dx = 1.2018\cdots,$

an underestimate with error $< 13^{-3} < 5 \cdot 10^{-4}$.

■ **Absolute Convergence:** If $\sum |a_n|$ converges, that is, if $\sum a_n$ {**converges absolutely**}, then $\sum a_n$ converges, and $\left| \sum_{n=1}^{\infty} a_n \right| \leq \sum_{n=1}^{\infty} |a_n|$.

◆ A series **converges conditionally** if it converges, but not absolutely.

■ **Comparison Tests:** Assume $a_n, b_n > 0$.

◆ If $\sum b_n$ converges and either $a_n \pounds b_n (n \geq N)$ or a_n / b_n has a limit, then $\sum a_n$ converges.

◆ If $\sum b_n$ diverges and either $b_n \pounds a_n (n \geq N)$ or a_n / b_n has a *nonzero* limit (or approaches ∞), then $\sum a_n$ diverges.

◆ The *p*-series and geometric series are often used for comparisons. Try a "limit" comparison when a series looks like a *p*-series, but is not directly comparable to it.

EX: $\displaystyle\sum_{n=K}^{\infty}\sin(1/n^2)$ converges since

$$\lim_{n\to\infty}\frac{\sin(1/n^2)}{1/n^2}=1.$$

Ratio & Root Tests: Assume $a_n\neq 0$.

◆ If $\displaystyle\lim_{n\to\infty}\frac{|a_{n+1}|}{a_n}<1$ or $\displaystyle\lim_{n\to\infty}|a_n|^{1/n}<1$, then $\displaystyle\sum a_n$ converges (**absolutely**).

◆ If $\displaystyle\lim_{n\to\infty}\frac{|a_{n+1}|}{a_n}>1$ or $\displaystyle\lim_{n\to\infty}|a_n|^{1/n}>1$, then $\displaystyle\sum a_n$ diverges.

◆ These tests are derived by comparison with geometric series. The following are useful in applying the root test:

$\displaystyle\lim_{n\to\infty}n^{p/n}=1$ (any p) and $\displaystyle\lim_{n\to\infty}(n!)^{1/n}=\infty$. More

precisely, $\displaystyle\lim_{n\to\infty}\frac{1}{n}(n!)^{1/n}=\frac{1}{e}$.

Power Series

Power Series: A power series in x is a sequence of

polynomials in x of the form $\displaystyle\sum_{n=0}^{N}a_n x^n\,(N=0,1,2,\ldots)$.

◆ The power series is denoted $\displaystyle\sum_{n=0}^{\infty}a_n x^n$.

◆ A power series in $x-c$ (or "centered at c" or

"about c") is written $\displaystyle\sum_{n=0}^{\infty}a_n(x-c)^n=a_0+$

$a_1(x-c)+a_2(x-c)^2+\cdots$.

◆ Replacing x with a real number q in a power series yields a series of real numbers. A power series **converges at q** if the the resulting series of real numbers converges.

■ **Interval of Convergence:** The set of real numbers at which a power series converges is an interval, called the **interval of convergence**, or a point.

◆ If the power series is centered at c, this set is either (i) $(-\infty, \infty)$; (ii) $(c-R, c+R)$ for some $R > 0$, possibly together with one or both endpoints; or (iii) the point c alone.

◆ In case (ii), R is called the **radius of convergence** of the power series, which may be ∞ and **0** for cases (i) and (iii), respectively. Convergence is absolute for $|x-c| < R$.

◆ You can often determine a radius of convergence by solving the inequality that puts the ratio (or root) test limit less than 1.

EX: For $\displaystyle\sum_{n=1}^{\infty} \frac{x^n}{2^n n^2}$, $\displaystyle\lim_{n\to\infty} \frac{|x|^{n+1}}{2^{n+1}(n+1)^2} \cdot \frac{2^n n^2}{|x|^n} \equiv$

$\dfrac{|x|}{2} < 1 \Rightarrow |x| < 2$, which, with the ratio test, shows that the radius of convergence is **2**.

■ **Geometric Power Series:** A power series determines a function on its interval of convergence: $x \mapsto f(x) = \displaystyle\sum_{n=0}^{\infty} a_n(x-c)^n$.

◆ One says the series converges to the function. The series $\displaystyle\sum_{n=0}^{\infty} x^n$, i.e., the sequence of polynomials

$$\sum_{n=0}^{N} x^n = 1 + x + x^2 + \ldots + x^N = \frac{1-x^{N+1}}{1-x} \, (x \neq 1),$$

converges for x in the interval $(-1, 1)$ to $1/(1-x)$ and diverges otherwise.

That is, $\sum_{n=0}^{\infty} x^n = \frac{1}{1-x} \, (\,|x| < 1)$. Other geometric series may be identified through this basic one.

EX: $\sum_{n=0}^{\infty} 2 \cdot 3^{-n} x^n = 2 \frac{x}{3} \sum_{n=0}^{\infty} \left(\frac{x}{3}\right)^n = \frac{2x}{3} \cdot \frac{1}{1-x/3}$, for $|x/3| < 1$. The interval of convergence is $(-3, 3)$.

■ **Calculus of Power Series:** Consider a function given by a power series centered at c with radius of convergence R: $f(x) = \sum_{n=0}^{\infty} a_n (x-c)^n$.

◆ Such a function is **differentiable** on $(c-R, c+R)$, and its derivative there is $f'(x) = \sum_{n=1}^{\infty} n a_n (x-c)^{n-1}$.

◆ The differentiated series has radius of convergence R, but may diverge at an endpoint where the original converged. Such a function is **integrable** on $(c-R, c+R)$, and its integral vanishing at c is:

$$\int_{c}^{x} f(t)\,dt = \sum_{n=0}^{\infty} \frac{a_n}{n+1} (x-c)^{n+1} \, (\,|x-c| < R).$$

◆ The integrated series has radius of convergence R, and may converge at an endpoint where the original diverged.

◆ The initial (geometric) series converges on $(-1,1)$, and the integrated series converges on $(1,-1)$. The integration says $\ln(1+x)=\sum\limits_{n=1}^{\infty}(-1)^{n+1}\dfrac{x^n}{n}$ for $|x|<1$; a remainder argument (see below) implies equality for $x=1$.

▨ **Taylor & MacLaurin Series:** The **Taylor series** about c of an infinitely differentiable function f is

$$\sum_{k=0}^{\infty}\frac{f^{(k)}(c)}{k!}(x-c)^k = f(c) + f'(c)(x-c) +$$

$$\frac{f''(c)}{2!}(x-c)^2+\cdots.$$

◆ If $c=0$, it is also called a **MacLaurin series**.
◆ The Taylor series at x may converge without converging to $f(x)$.
◆ It converges to $f(x)$ if the remainder in Taylor's formula, $R_n(x)=\dfrac{1}{(n+1)!}f^{(n+1)}(\xi)\cdot(x-c)^{(n+1)}$ (ξ between c and x, ξ varying with x and n), approaches 0 as $n\to\infty$.

EX: The remainders at $x=1$ for the MacLaurin polynomials of $\ln(1+x)$ (in **Taylor's formula** above) satisfy $\left|R_n(1)\right|=\dfrac{1}{(n+1)(1+\xi)^{n+1}}\le$ $\dfrac{1}{n+1}\to 0$, so $\ln 2 = \sum\limits_{n=1}^{\infty}\dfrac{(-1)^{n+1}}{n}$.

■ **Computing Taylor Series:** If $R>0$ and $f(x)=$ $\sum_{n=0}^{\infty} a_n(x-c)^n (|x-c|<R)$, then the coefficients are necessarily the Taylor coefficients: $a_n=f^{(n)}(c)/n!$.

◆ This means Taylor series may be found other than by directly computing coefficients.

◆ Differentiating the geometric series gives

$$\frac{1}{(1-x)^2} = \sum_{n=1}^{\infty} nx^{n-1} = \sum_{n=0}^{\infty} (n+1)x^n (|x|<1).$$

■ **Basic MacLaurin Series:**

$$\frac{1}{1-x}=1+x+x^2+...= \sum_{n=0}^{\infty} x^n \quad (|x|<1)$$

$$\ln(1+x)=x-\frac{x^2}{2}+\frac{x^3}{3}-...= \sum_{n=1}^{\infty} (-1)^{(n+1)} \frac{x^n}{n} (-1<x\leq 1)$$

$$\ln\frac{1+x}{1-x}=2\left(x+\frac{x^3}{3}+\frac{x^5}{5}+\cdots\right)=2 \sum_{n=0}^{\infty} \frac{x^{2n+1}}{2n+1} (|x|<1)$$

$$\arctan x=x-\frac{x^3}{3}+\frac{x^5}{5}-\cdots= \sum_{n=0}^{\infty} (-1)^n \frac{x^{2n+1}}{2n+1} (|x|\leq 1)$$

◆ The following hold for all real x:

$$e^x=1+x+\frac{x^2}{2!}+\frac{x^3}{3!}+\cdots= \sum_{n=0}^{\infty} \frac{x^n}{n!}$$

$$\cos x=1-x+\frac{x^2}{2!}+\frac{x^4}{4!}-\cdots= \sum_{n=0}^{\infty} \frac{(-1)^n x^{2n}}{(2n)!}$$

$$\sin x=x-\frac{x^3}{3!}+\frac{x^5}{5!}-\cdots= \sum_{n=0}^{\infty} \frac{(-1)^n x^{2n+1}}{(2n+1)!}$$

■ **Binomial Series:** For $p \neq 0$, and for $|x| < 1$,

$$(1+x)^p = 1 + px + \frac{p(p-1)}{2!}x^2 + \cdots = \sum_{n=0}^{\infty} \binom{p}{k} x^k.$$

◆ The **binomial coefficients** are $\binom{p}{0} = 1$, $\binom{p}{1} = p$,

$\binom{p}{2} = \frac{p(p-1)}{2}$, and ("$p$ choose k")

$$\binom{p}{k} = \frac{p(p-1)(p-2)\cdots(p-k+1)}{k!}$$

◆ If p is a positive integer, $\binom{p}{k} = 0$ for $k > p$.

11 Limits & Continuity

NOTES
This chapter provides the **methods** and **formulas** for *limits* and *continuity*.

■ $\lim_{x \to a} f(x) = L$ if $f(x)$ is **close to** L for all x sufficiently close (but not equal) to a.

■ $f(x)$ is continuous at $x = a$ if:
  ◆ $f(a)$ is defined,
  ◆ $\lim_{x \to a} f(x) = L$ exists, and
  ◆ $L = f(a)$

12 Integrals

The Definite Integral

- ### Let $f(x)$ Be Continuous on $[a, b]$

  - ### Riemann Sum Definition of Definite Integral

    - Divide $[a, b]$ into n equal subintervals of length
      $$h = \frac{b-a}{n}$$

    - Let $x_0 = a$, x_1, x_2, ..., $x_n = b$ denote the endpoints of the subintervals. They are found by: $x_0 = a$, $x_1 = a + h$, $x_2 = a + 2h$, $x_3 = a + 3h$, ..., $x_n = a + nh = b$.

    - Let m_1, m_2, ...m_n denote the midpoints of the subintervals. They are found by:
      $m_1 = 0.5(x_0 + x_1)$, $m_2 = 0.5(x_1 + x_2)$,
      $m_3 = 0.5(x_2 + x_3)$, ..., $m_n = 0.5(x_n - 1 + x_n)$.

      $$\int_a^b f(x)dx \approx h[f(m_1) + f(m_2) + ... + f(m_n)]$$

  - ### Midpoint Rule:

    $$\int_a^b f(x)dx = \lim_{n \to \infty} h[f(x_1) + f(x_2) + ... + f(x_n)]$$

  - ### Trapezoid Rule:

    $$\int_a^b f(x)dx \approx \frac{h}{2}[f(x_0) + 2f(x_1) + 2f(x_2) + ... + 2f(x_{n-1}) + f(x_n)]$$

◆ **Simpson's Rule:**

$$\int_a^b f(x)dx \approx \frac{h}{6}[f(x_0) + 4f(m_1) + 2f(x_1) +$$
$$4f(m_2) + 2f(x_2) + ... + 2f(x_{n-1}) + 4f(m_n) + f(x_n)]$$

The Indefinite Integral

■ **$F(x)$ Is Called an Antiderivative of $f(x)$, If $F'(x) = f(x)$**

◆ The most general antiderivative is denoted $\int f(x)dx$

◆ $\int f(x)dx$ is also called the Indefinite Integral of $f(x)$

■ **Fundamental Theorem of Calculus**

◆ If $F'(x) = f(x)$ and $f(x)$ is continuous on $[a, b]$, then $\int_a^b f(x)dx = F(b) - F(a)$

Integration Formulas

■ $\int [f(x) + g(x)]dx = \int f(x)dx + \int g(x)dx$

■ $\int kf(x)dx = k \int f(x)dx$ if k is a constant

■ $\int u^n\, du = \dfrac{u^{n+1}}{n+1} + C$

■ $\int \dfrac{1}{u}\, du = \ln|u| + C$

■ $\int e^u\, du = e^u + C$

■ If $y = f(x) \geq 0$ on $[a, b]$, $\int_a^b f(x)dx$ gives the area under the curve.

■ If $f(x) \geq g(x)$ on $[a, b]$, $\int_a^b [f(x) - g(x)]dx$ gives the area between the 2 curves $f(x)$ and $g(x)$

■ Average value of $f(x)$ on $[a, b]$ is $\frac{1}{b-a}\int_a^b f(x)dx$

■ Volume of the solid of revolution obtained by revolving about the x-axis the region under the curve $y = f(x)$ from $x = a$ to $x = b$ is $\int_a^b \pi[f(x)]^2 dx$

Integration by Parts

■ Factor the integrand into 2 parts: u and dv

■ Find du and $v = \int dv$

■ Find $\int vdu$

■ Set $\int udv = uv - \int vdu$

Integration by Substitution

■ **To Solve** $\int f(g(x))g'(x)dx$

 ◆ Set $u = g(x)$, where $g(x)$ is chosen so as to simplify the integrand

 ◆ Substitute $u = g(x)$ and $du = g'(x)dx$ into the integrand.

   • This step usually requires multiplying or dividing by a constant.

 ◆ Solve $\int f(u)du = F(u) + C$.

 ◆ Substitute $u = g(x)$ to get the answer: $F(g(x))+C$.

Improper Integrals

■ Infinite Limits of Integration

◆ $\int_a^\infty f(x)dx = \lim\limits_{b\to\infty} \int_a^b f(x)dx$

◆ $\int_{-\infty}^b f(x)dx = \lim\limits_{a\to-\infty} \int_a^b f(x)dx$

■ Improper at the Left or Right Endpoints

◆ If $f(x)$ is discontinuous at $x = b$,

$$\int_a^b f(x)dx = \lim\limits_{h\to b^-} \int_a^h f(x)dx$$

◆ If $f(x)$ is discontinuous at $x = a$,

$$\int_a^b f(x)dx = \lim\limits_{h\to\infty} \int_a^h f(x)dx$$

13 Derivatives & Their Applications

> **NOTES**
> This chapter provides the **methods** for applications of *derivatives* and *differentials*.

Derivative Basics

Definition of Derivative

♦ $f'(a) = \lim\limits_{h \to 0} \dfrac{f(a+h)-f(a)}{h}$

♦ If $y = f(x)$, the derivative $f'(x)$ is also denoted $\dfrac{dy}{dx}$

Formulas:

♦ **Power Rule:** $\dfrac{d}{dx}(x^n) = nx^{n-1}$

♦ $\dfrac{d}{dx}(e^{kx}) = ke^{kx}$

♦ $\dfrac{d}{dx}(\ln x) = \dfrac{1}{x}$

♦ **General Power Rule:**

$\dfrac{d}{dx}([f(x)]^n) = n[f(x)]^{n-1}f'(x)$

♦ $\dfrac{d}{dx}[e^{f(x)}] = e^{f(x)}f'(x)$

♦ $\dfrac{d}{dx}[\ln f(x)] = \dfrac{f'(x)}{f(x)}$

♦ **Sum or Difference Rule:**

$\dfrac{d}{dx}[f(x) \pm g(x)] = f'(x) \pm g'(x)$

◆ **Constant Multiple Rule:**

$$\frac{d}{dx}[kf(x)] = kf'(x)$$

◆ **Product Rule:**

$$\frac{d}{dx}[f(x)g(x)] = f'(x)g(x) + f(x)g'(x)$$

◆ **Quotient Rule:**

$$\frac{d}{dx}\left[\frac{f(x)}{g(x)}\right] = \frac{f'(x)g(x) - f(x)g'(x)}{[g(x)]^2}$$

◆ **Chain Rule:** $\dfrac{d}{dx}[f(g(x))] = f'(g(x))g'(x),$ or

$$\frac{dy}{dx} = \frac{dy}{du} \cdot \frac{du}{dx}$$

◆ **Derivative of an Inverse Function:** $\dfrac{dx}{dy} = \dfrac{1}{\frac{dy}{dx}}$

Implicit Differentiation

■ **Given an Equation Involving Function of x & y,**

to Find: $\dfrac{dy}{dx}$

◆ Differentiate both sides of the equation with respect to x, treating y as a function of x and applying the chain rule to each term involving y

i.e. $\dfrac{d}{dx}[f(y)] = f'(y)\dfrac{dy}{dx}$.

◆ Move all terms with $\dfrac{dy}{dx}$ to left side and all other terms to the right.

◆ Solve for $\dfrac{dy}{dx}$.

Curve Sketching
■ Steps to Follow in Sketching the Curve, $y = f(x)$:
◆ Determine the domain of $f(x)$.
◆ Analyze all points where $f(x)$ is discontinuous. Sketch the graph near all such points.
◆ Test for vertical, horizontal and oblique asymptotes.
  • $f(x)$ has a vertical asymptote at $x = a$ if:
$$\lim_{x \to a^-} f(x) = \pm\infty \text{ or } \lim_{x \to a^+} f(x) = \pm\infty$$
  • $f = (x)$ has a horizontal asymptote $y = b$ if:
$$\lim_{x \to \infty} f(x) = b \text{ or } \lim_{x \to -\infty} f(x) = b$$
  • Sketch any asymptotes.
◆ Find $f'(x)$ and $f''(x)$.
◆ Find all critical points. These are points $x=a$ where $f'(a)$ does not exist or $f'(a)=0$. Repeat the steps that follow for each critical point $x = a$:
  • If $f(x)$ is continuous at $x = a$,
    > $f(x)$ has a relative maximum at $x = a$ if:
      - $f'(a) = 0$ and $f''(a) < 0$, or
      - $f'(x) > 0$ to the left of a and $f'(x) < 0$ to the right of a.
    > $f(x)$ has a relative minimum at $x=a$ if:
      - $f'(a) = 0$ and $f''(a) > 0$, or
      - $f'(x) < 0$ to the left of a and $f'(x) > 0$ to the right of a.
  • Sketch $f(x)$ near $(a, f(a))$.
◆ Find all possible inflection points. These are points $x = a$ where $f''(x)$ does not exist or $f''(x)=0$. Repeat the steps that follow for each such $x=a$:
  • $f(x)$ has an inflection point at $x = a$ if $f(x)$ is continuous at $x = a$ and

> $f''(x) < 0$ to the left of a and $f''(x) > 0$ to the right of a, **or**
> $f''(x) > 0$ to the left of a and $f''(x) < 0$ to the right of a.
* Sketch $f(x)$ near $(a, f(a))$.
◆ If possible, plot the x- and y- intercepts.
◆ Finish the sketch.

Optimization Problems

■ **To Optimize Some Quantity Subject to Some Constraint:**
  ◆ Identify and label quantity to be maximized or minimized.
  ◆ Identify and label all other quantities.
  ◆ Write quantity to be optimized as a function of the other variables. This is called the objective function (or objective equation).
  ◆ If the objective function is a function of more than 1 variable, find a constraint equation relating the other variables.
  ◆ Use the constraint equation to write the objective function as a function of only 1 variable.
  ◆ Using the curve sketching techniques, locate the maximum or minimum of the objective function.

Approximations & Differentials

■ **Let $y = f(x)$ & Assume $f'(a)$ Exists**
  ◆ The Equation of the Tangent Line to $y = f(x)$ at the point $(a, f(a))$ is $y - f(a) = f'(a)(x-a)$
  ◆ The differential of y is $dy = f'(x)dx$
  ◆ Linear Approximation, or Approximation by Differentials:
    • Set $dx = \Delta x = x - a$, $\Delta y = f(x) - f(a)$

• The equation of the tangent line becomes:
$\Delta y = f'(a)\Delta x = f'(a)dx$. If Δx is small, then
$\Delta y \approx dy$. That is, $f(x) \approx f(a) + f'(a)(x - a)$.

◆ The nth Taylor polynomial of $f(x)$ centered at
$x = a$ is $p_n(x) = f(a) + \dfrac{f'(a)(x-a)}{1!} +$

$\dfrac{f''(a)(x-a)^2}{2!} + ... + \dfrac{f^{(n)}(a)(x-a)^n}{n!}$

Motion
■ Formula

◆ If $s = s(t)$ represents the position of an object at
time t relative to some fixed point, then
$v(t)=s'(t) =$ velocity at time t and $a(t) = v'(t) =$
$s''(t) =$ acceleration at time t.

Applications to Business & Economics

14

NOTES
This chapter outlines various **methods** that are essential to *business* and *economics*.

Cost, Revenue & Profit

- $C(x)$ = cost of producing x units of a product
- $p = p(x)$ = price per unit; ($p = p(x)$ is also called the demand equation)
- $R(x) = xp$ = revenue made by producing x units
- $P(x) = R(x) - C(x)$ = profit made by producing x units
- $C'(x)$ = marginal cost
- $R'(x)$ = marginal revenue
- $P'(x)$ = marginal profit

Compounding Interest

- **Starting with a Principal p_0**

  ◆ If the interest is compounded for t years with m periods per year at the interest rate of r per annum, the compounded amount is:

  $$P = P_0(1 + \frac{r}{m})^{mt}.$$

  ◆ If interest is continuously compounded, $m \rightarrow \infty$ and the formula becomes:

  $$P = \lim_{m \to \infty} P_0(1 + \frac{r}{m})^{mt} = P = P_0 e^{rt}.$$

◆ The formula $P = P_0 e^{rt}$ gives the value at the end of t years, assuming continuously compounded interest. P_0 is called the present value of P to be received in t years and is given by the formula: $P_0 = Pe^{-rt}$.

Elasticity of Demand

■ **Solving for x in the Demand Equation $p = p(x)$ Gives $x = f(p)$**

◆ Demand function which gives the quantity demanded x as a function of the price p.

◆ The elasticity of demand is: $E(p) = \dfrac{-pf'(p)}{f(p)}$

■ **Demand Is Elastic at $p = p0$ if $E(p0) > 1$**

In this case an increase in price corresponds to a decrease in revenue.

■ **Demand Is Inelastic at $p = p0$ if $E(p0) < 1$**

In this case an increase in price corresponds to an increase in revenue.

Consumers' Surplus

■ **If a Commodity Has Demand Equation $p = p(x)$**

Consumers' Surplus is given by $\int_0^a [p(x) - p(a)] dx$

where a is the quantity demanded and $p(a)$ is the corresponding price.

15

Exponential Models

Exponential Growth & Decay

■ **Exponential Growth:** $y = P_0 e^{kt}$

◆ Satisfies the differential equation $y' = ky$

◆ P_0 is the initial size, $k > 0$ is called the growth constant.

◆ The time it takes for the size to double is given by: $\frac{\ln 2}{k}$.

■ **Exponential Decay:** $y = p_0 e^{-\lambda t}$

◆ Satisfies the differential equation $y' = -\lambda y$

◆ P_0 is the initial size, $\lambda > 0$ is called the decay constant.

◆ The half life $t_{1/2}$ is the time it takes for y to become $P_0/2$. It is found by $t_{1/2} = \frac{\ln .5}{-\lambda} = \frac{\ln 2}{-\lambda}$.

Other Growth Curves

■ **The Learning Curve:** $y = m(1 - e^{-kt})$

Satisfies the differential equation $y' = k(M - y)$, $y(0) = 0$ where M and k are positive constants.

■ The Logistic Growth Curve:

$y = \dfrac{M}{1 + Be^{-Mkt}}$ satisfies the differential equation

$y' = ky(M-y)$ where B, M and k are positive constants.

16 Probability

Definitions

■ Probability Density Function

For the continuous random variable x is a function $p(x)$ satisfying: and $p(x) \geq 0$ if $A \leq x \leq B$ and $\int_A^B p(x)dx = 1$, where we assume the values of x lie in $[A, B]$.

■ The Probability That

$a \leq x \leq b$ is $P[a \leq x \leq b] = \int_a^b p(x)dx$

■ Expected Value (or Mean of X)

Given by $m = E(X) = \int_A^B xp(x)dx$

■ Variance of X: Given by $\sigma^2 = var(X) = $
$\int_A^B (x-\mu)^2 p(x)dx = \int_A^B x^2 p(x)dx - \mu^2$

Common Probability Density Functions

■ Uniform Distribution Function:

$$p(x) = \frac{1}{B-A}, \mu = E(x) = \frac{B+A}{2}, var(x) = \frac{(B-A)^2}{12}$$

■ **Exponential Density Function:**

$p(x) = \lambda e^{-\lambda x}$. In this case, $A = 0$, $B = \infty$,

$\mu = E(X) = 1/\lambda$, $var(X) = 1/\lambda^2$.

■ **Normal Density Function:**

with $E(X) = \mu$ and $var(X) = \sigma^2$ is:

$$p(x)\frac{1}{\sqrt{2\pi}\,\sigma}e^{-\frac{(x-\mu)^2}{2\sigma^2}} = \frac{1}{\sqrt{2\pi}\,\sigma}exp\left[\frac{(x-\mu)^2}{2s^2}\right]$$

17 Calculus of Functions of Two Variables

> **NOTES**
> This chapter outlines **methods** for applications of *functions with more than one variable.*

Partial Derivatives

Where $f(x, y)$ Is a Function of Two Variables x & y

◆ $\dfrac{\partial f}{\partial x}$ is the derivative of $f(x, y)$ with respect to x, treating y as a constant.

◆ $\dfrac{\partial f}{\partial y}$ is the derivative of $f(x, y)$ with respect to y, treating x as a constant.

◆ $\dfrac{\partial^2 f}{\partial x^2} = \dfrac{\partial}{\partial y} - \dfrac{\partial f}{\partial x}$ is the second partial derivative of $f(x, y)$ with respect to x twice, keeping y constant each time.

◆ $\dfrac{\partial^2 f}{\partial x \partial y} = \dfrac{\partial}{\partial y} \dfrac{\partial f}{\partial x}$ is the second partial derivative of $f(x, y)$ first with respect to x keeping y constant then with respect to y keeping x constant.

◆ Other notation for partial derivatives:

$$f_x(x, y) = \frac{\partial f}{\partial x}, f_{xx}(x, y) = \frac{\partial^2 f}{\partial x^2}, f_{xy}(x, y) = \frac{\partial^2 f}{\partial y \partial x}$$

Differentials

■ **If $f = f(x, y)$**

♦ $df = \dfrac{\partial f}{\partial x}dx + \dfrac{\partial f}{\partial y}dy = f_x(x, y)dx + f_y(x, y)dy$

♦ Setting $dx = \Delta x = x - a$, $dy = \Delta y = y - b$ and $\Delta f = f(x, y) - f(a, b)$, if Δx and Δy are both small, then $\Delta f \approx df$.
That is: $f(x, y) \approx f(a, b) + f_x(a, b)\Delta x + f_y(a, b)\Delta y$.

Relative Extrema Test

■ **To Locate Relative Maxima, Relative Minima & Saddle Points on the Graph of $z = f(x, y)$.**

♦ Solve simultaneously: $\dfrac{\partial f}{\partial x} = 0$ and $\dfrac{\partial f}{\partial y} = 0$. For each ordered pair (a, b) such that $\dfrac{\partial f}{\partial x}(a, b) = 0$ and $\dfrac{\partial f}{\partial y}(a, b) = 0$, apply the following test.

♦ Set $A = \dfrac{\partial^2 f}{\partial x^2}(a, b)$, $B = \dfrac{\partial^2 f}{\partial y^2}(a, b)$, $C = \dfrac{\partial^2 f}{\partial x \partial y}(a, b)$ and $D = AB - C^2$

• If $D > 0$ and $A > 0$, then $f(x, y)$ has a relative minimum at (a, b).
• If $D > 0$ and $A < 0$, then $f(x, y)$ has a relative maximum at (a, b).
• If $D < 0$, then $f(x, y)$ has a saddle point at (a, b).
• If $D = 0$, then the test fails. $f(x, y)$ may or may not have an extremum or saddle point at (a, b).

The Method of Lagrange Multipliers

■ **Solving Constrained Optimization Problems to Maximize or Minimize $f(x, y)$, Subject to the Constraint $g(x, y) = 0$**

◆ Define the new function $F(x, y, \lambda) = f(x, y) + \lambda g(x, y)$.

◆ Solve the system of 3 equations:

• $\dfrac{\partial F}{\partial x} = 0$,

• $\dfrac{\partial F}{\partial y} = 0$, and

• $\dfrac{\partial F}{\partial \lambda} = 0$ simultaneously.

> **This is usually accomplished in 4 steps:**

- **Step 1:** Solve $\dfrac{\partial F}{\partial x} = 0$, and $\dfrac{\partial F}{\partial y} = 0$, for λ and equate the solutions.

- **Step 2:** Solve the resulting equation for one of the variables, x or y.

- **Step 3:** Substitute this expression for x or y into equation $\dfrac{\partial F}{\partial \lambda} = 0$. and solve the resulting equation of one variable for the other variable.

- **Step 4:** Substitute the value found in Step 3 into the equation found in Step 2. Use 1 of the equations from Step 1 to find λ. This gives the value of x and y.

Double Integrals

▮ Formulas

◆ If **R** is the region in the plane bounded by the 2
curves $y = g(x)$, $y = h(x)$ and the 2 vertical lines
$x = a$, $x = b$, then the double integral
$\iint\limits_{R} f(x,y)\,dx\,dy$ is equal to the iterated integral
$\int_{a}^{b}\left(\int_{g(x)}^{h(x)} f(x,y)\,dy\right)dx.$

◆ To   evaluate   the   iterated   integral
$I = \int_{a}^{b}\left(\int_{g(x)}^{h(x)} f(x,y)\,dy\right)dx$

• Find an antiderivative $F(x,y)$ for $f(x,y)$ with
respect to y keeping x constant.

That is: $\dfrac{\partial F}{\partial y} = f(x,y)$

• Set: $I = \int_{a}^{b}[F(x,h(x)) - F(x,g(x))]\,dx$

• Solve this integral. The integrand is a function
of one variable.

Differential Equations

> **NOTES**
> This chapter provides **methods** for applications
> of *differential equations*.

Definitions

- **A Differential Equation (DE) Is:** Any equation
  involving a derivative. For example, it could be an
  equation involving $\frac{dy}{dx}$ (*or y'*, or *y'(x)*), *y* and *x*.

- **A Solution Is**: A function $y = y(x)$ such that $\frac{dy}{dx}$, *y*
  and *x* satisfy the original equation.

- **An Initial Value Problem (IVP):** Also specifies
  the value of the solution *y(a)* at some point *x = a*

- **Simple Differential Equations (DEs):** Can be
  solved by separation of variables and integration.

  ◆ For example, the equation $f(x) = g(y)\frac{dy}{dx}$ can be
    written as $f(x)dx = g(y)dy$ and can be solved by
    integrating both sides: $\int f(x) = \int g(y)\frac{dy}{dx}$.

19 Formulas from Pre-Calculus

Logarithms & Exponentials

- $y = \ln x$ if and only if $x = e^y = \exp(y)$
- $\ln e^x = x$
- $e^{\ln x} = x$
- $e^x e^y = e^{x+y}$
- $\dfrac{e^x}{e^y} = e^{x-y}$
- $(e^x)^y = e^{xy}$
- $e^0 = 1$
- $\ln(xy) = \ln x + \ln y$
- $\ln(x/y) = \ln x - \ln y$
- $\ln(x^y) = y \ln x$
- $\ln 1 = 0$
- $\ln e = 1$

Algebraic Formulas

- If $a \neq 0$, the solutions to $ax^2 + bx + c = 0$ are given by $x = \dfrac{-b \pm \sqrt{b^2 - 4ac}}{2a}$.
- Point-slope equation of a line: $y - y_0 = m(x - x_0)$

20 Geometric Formulas

- **Area:** The area, **A**, of a 2-dimensional shape is the number of square units that can be put in the region enclosed by the sides.
  - ◆ Area is obtained through some combination of multiplying heights and bases, which always form 90° angles with each other, except in circles.
- **Perimeter:** The perimeter, **P**, of a 2-dimensional shape is the sum of all side lengths.
- **Volume:** The volume, **V**, of a 3-dimensional shape is the number of cubic units that can be put in the space enclosed by all the sides.

Square Area: $A = b^2$
If $b = 8$, then:
$A = 64$ square units

Rectangle Area: $A = hb$, or $A = lw$
If $h = 4$ and $b = 12$ then:
$A = (4)(12)$, $A = 48$ square units

Triangle Area: $A = \frac{1}{2}bh$
If $h = 8$ and $b = 12$ then:
$A = \frac{1}{2}(8)(12)$, $A = 48$ square units

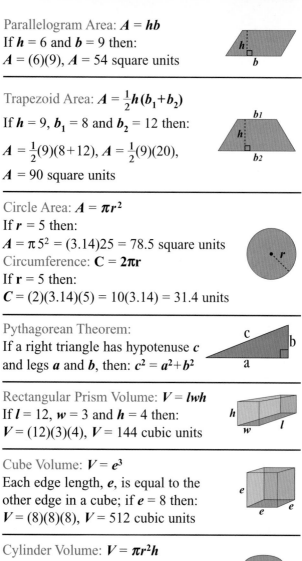

Parallelogram Area: $A = hb$
If $h = 6$ and $b = 9$ then:
$A = (6)(9)$, $A = 54$ square units

Trapezoid Area: $A = \frac{1}{2}h(b_1 + b_2)$
If $h = 9$, $b_1 = 8$ and $b_2 = 12$ then:
$A = \frac{1}{2}(9)(8+12)$, $A = \frac{1}{2}(9)(20)$,
$A = 90$ square units

Circle Area: $A = \pi r^2$
If $r = 5$ then:
$A = \pi 5^2 = (3.14)25 = 78.5$ square units
Circumference: $C = 2\pi r$
If $r = 5$ then:
$C = (2)(3.14)(5) = 10(3.14) = 31.4$ units

Pythagorean Theorem:
If a right triangle has hypotenuse c
and legs a and b, then: $c^2 = a^2 + b^2$

Rectangular Prism Volume: $V = lwh$
If $l = 12$, $w = 3$ and $h = 4$ then:
$V = (12)(3)(4)$, $V = 144$ cubic units

Cube Volume: $V = e^3$
Each edge length, e, is equal to the
other edge in a cube; if $e = 8$ then:
$V = (8)(8)(8)$, $V = 512$ cubic units

Cylinder Volume: $V = \pi r^2 h$
If radius $r = 9$ and $h = 8$ then:
$V = \pi(9)^2(8)$, $V = (3.14)(81)(8)$,
$V = 2034.72$ cubic units

Cone Volume: $V = \frac{1}{3}\pi r^2 h$

If $r = 6$ and $h = 8$ then:

$V = \frac{1}{3}\pi(6)^2(8)$, $V = \frac{1}{3}(3.14)(36)(8)$,

$V = 301.44$ cubic units

Triangular Prism Volume: $V = \textbf{(area of triangle)}h$

If has an area equal to $\frac{1}{2}(5)(12)$ then:

$V = 30h$ and if $h = 8$ then:
$V = (30)(8)$, $V = 240$ cubic units

Rectangular Pyramid Volume:

$V = \frac{1}{3}\textbf{(area of rectangle)}h$

If $l = 5$ and $w = 4$, the rectangle has an area of 20, then:

$V = \frac{1}{3}(20)h$ and if $h = 9$ then:

$V = \frac{1}{3}(20)(9)$, $V = 60$ cubic units

Sphere Volume: $V = \frac{4}{3}\pi r^3$

If radius $r = 5$ then:

$V = \frac{4}{3}(3.14)(5)^3$, $V = 523.3$ cubic units

21 Develop a Problem Solving Strategy

NOTES
- **3** Key Issues
- **8** Useful Steps

■ **3 Key Issues:**
- ◆ Understand the business or scientific **principle** required to solve the problem.
- ◆ Develop a correct mathematical **strategy**.
- ◆ Logically **approach** solving the problem.

■ **8 Useful Steps in Problem Solving:**
- ◆ **Prepare a rough sketch** or diagram based on the subject of the problem. For business applications, set this up using business terms; for science, use physical variables.
- ◆ **Identify all relevant variables**, concepts and constants.
  Note: Do not simply search for the "right" equation in your notes or text. You may have to select your own variables to solve the problem.
- ◆ **Describe the problem** using appropriate mathematical relationships or graphs.
- ◆ **Obtain any constants** from the stated problem or textbook. Make sure you have all the **essential data**.
  Hint: You may have extra information.

◆ **The hard part:** Derive a **mathematical expression** for the problem. Make sure that the equation, constants and data give the right unit for the final answer.

◆ **Carry out the appropriate mathematical manipulation**, differentiate, integrate, find limits, etc.

◆ **The easy part**: plug numbers into the equation. Obtain a quick **approximate answer**, then use a calculator to obtain an exact numerical answer.

◆ **Check the final answer** using the original statement of the problem, your sketch and common sense. Are the units, sign, magnitude all correct?

Glossary

■ **Absolute value** of a real number x (written $|x|$) is the distance that the number is from zero.

■ **Algebraic function** $y = f(x)$ is a function that satisfies the equation: $P_n y^n + P_{n-1} y^{n-1} + ... + P_1 y + P_0 = 0$ (where P is a polynomial in x with rational coefficients).

■ **Antiderivative** $F(x)$ of a function f is a function such that $F'(x) = f(x)$ for all x in the domain of f.

■ **Arithmetic sequence** is a sequence in which the difference d between any two consecutive terms is constant.

■ **Continuity** at an interior point $x = c$ of the domain of function f exists if $\lim_{x \to c} f(x) = f(c)$.

■ **Decreasing function** f is a function that is defined on an interval with x_1 and x_2 being any two points in the interval such that f decreases on the interval if $x_1 < x_2$ implies $f(x_1) > f(x_2)$.

■ **Dependent variable** is the output variable in an equation and depends on or is determined by the input variable.

■ **Derivative** of a function f with respect to the variable x is the function f' that has the value at x of $f'(x) = \lim_{h \to 0} \dfrac{f(x+h)-f(x)}{h}$, provided the limit exists.

■ **Domain** of a function is the set of input values.

■ **Even function** $f(x)$ is an even function if $f(-x) = f(x)$ for every number x in the domain of $f(x)$.

■ **Exponential function** $f(x)$ is an exponential function if it involves or includes b^x where the base is b and $b > 0$ and $b \neq 1$.

■ **Function** is a relation with exactly one output for each input.

■ **Geometric sequence** is a sequence that has a constant ratio r of any term to the consecutively previous term.

■ **Hyperbolic function** is a function that is defined on an interval centered at the origin and can be written in a unique way as the sum of one even function and one odd function with the decomposition of $f(x) = \underbrace{\dfrac{f(x)+f(-x)}{2}}_{even} + \underbrace{\dfrac{f(x)-f(-x)}{2}}_{odd}$

or $e^x = \underbrace{\dfrac{e^x + e^{-x}}{2}}_{even} + \underbrace{\dfrac{e^x - e^{-x}}{2}}_{odd}$.

■ **Increasing function** f is a function that is defined on an interval with x_1 and x_2 being any two points in the interval such that f increases on the interval if $x_1 < x_2$ implies $f(x_1) < f(x_2)$.

■ **Indefinite integral** is the set of all antiderivatives of f with respect to x, denoted by $\int f(x)dx$.

■ **Independent variable** is the input variable in an equation.

■ **Inverse function** g is a function that maps the output values of the original function f back to their original input values of f such that $f(g(x)) = x$ and $g(f(x)) = x$; the graph of the inverse function g is the reflection of the original function f over the line $y = x$.

■ **Limit** of $f(x)$ is L, written $\lim_{x \to c} f(x) = L$, if $f(x)$ is defined on an open interval about c, except possibly at c itself, and $f(x)$ gets arbitrarily close to L for all values of x sufficiently close to c; f approaches the limit L as x approaches c.

■ **Logarithm** of positive number y with base b, where $b > 0$ and $b \neq 1$, is equal to x (written $\log_b y = x$) if and only if $b^x = y$.

■ **Odd function** $f(x)$ is an odd function if $f(-x) = -f(x)$ for every number x in the domain of $f(x)$.

■ **Radian** (one radian) is the measure of a central angle of a circle in standard position whose terminal side intercepts an arc that has the same length as the radius of the circle.

■ **Range** of a function is the set of output values.

■ **Rational function** is a function of the form $f(x) = \dfrac{p(x)}{q(x)}$ where $p(x)$ and $q(x)$ are polynomials and $q(x) \neq 0$.

■ **Sequence** of numbers is a function whose domain is the set of integers greater than or equal to some integer x_0.

■ **Series** is the expression that results when the terms of a sequence are added.